Engineering the Channel Tunnel

Engineering the Channel Tunnel

EDITED BY
Colin J. Kirkland

E & FN SPON
An Imprint of Chapman & Hall

London · Glasgow · Weinheim · New York · Tokyo · Melbourne · Madras

Co-published by E & FN Spon, an imprint of Chapman & Hall, 2–6 Boundary Row, London, SE1 8HN and James & James (Science Publishers) Ltd, Waterside House, 47 Kentish Town Road, London NW1 8NZ

Chapman & Hall, 2–6 Boundary Row, London SE1 8HN, UK

Blackie Academic & Professional, Wester Cleddens Road, Bishopbriggs, Glasgow G64 2NZ, UK

Chapman & Hall GmbH, Pappelallee 3, 69469 Weinheim, Germany

Chapman & Hall USA, 115 Fifth Avenue, New York, NY 10003, USA

Chapman & Hall Japan, ITP-Japan, Kyowa Building 3F, 2–2–1 Hirakawacho, Chiyado-ku, Tokyo 102, Japan

Chapman & Hall Australia, 102 Dodds Street, South Melbourne, Victoria 3205, Australia

Chapman & Hall India, R. Seshadri, 32 Second Main Road, CIT East, Madras 600 035, India

---

First edition 1995

© 1995 Eurotunnel

Typeset in 10/12 Palatino
Printed in Great Britain by St Edmondsbury Press, Bury St Edmunds, Suffolk

ISBN 0 419 17920 8

Apart from any fair dealing for the purposes of research or private study, or criticism or review, as permitted under the UK Copyright Designs and Patents Act, 1988, this publication may not be reproduced, stored, or transmitted, in any form or by any means, without the prior permission in writing of the publishers, or in the case of reprographic reproduction only in accordance with the terms of the licences issued by the Copyright Licensing Agency in the UK, or in accordance with the terms of licences issued by the appropriate Reproduction Rights Organization outside the UK. Enquiries concerning reproduction outside the terms stated here should be sent to the publishers at the London address printed on thispage.
 The publisher makes no representation, express or implied, with regard to the accuracy of the information contained in this book and cannot accept any legal responsibility or liability for any errors or omissions that may be made.

A catalogue record of this book is available from the British Library

∞ Printed on acid-free text paper, manufactured in accordance with ANSI/NISO Z39, 48-1992 (Permanence of Paper).

Contents

Acknowledgements   vii

1   Introduction
      COLIN J KIRKLAND   1

Section I: Survey/Geology

2   Geology
      COLIN WARREN AND PAUL VARLEY   21
3   Control and construction surveys
      ERIC RADCLIFFE   51

Section II: Tunnelling

4   Tunnel lining design
      GUY LANCE   63
5   Initial access developments
      DAVID WALLIS   79
6   Tunnel boring machines
      BOB MARSHALL   91
7   Construction logistics
      TIM GREEN   117

Section III: Railway/tunnel services

Railway design
8   Radio links
      KAREL DE JAEGER-PONNET   129
9   Power system
      JOHN FINN   133
10   Signalling and train control
      PAUL ROBINS   149
11   Terminal and tunnel trackwork
      MICHAEL BAXTER AND PETER DAVIES   157

Rolling stock

12 Locomotives
    ROGER FORD 175
13 Passenger-vehicle shuttle fleet
    PETER SEMMENS AND YVES MACHEFERT-TASSIN 191
14 Freight shuttle fleet
    PETER SEMMENS AND YVES MACHEFERT-TASSIN 203

Tunnel services

15 Aerodynamics, ventilation and cooling
    DAVID HENSON 211
16 Fire detection and suppression
    ERIC H WHITAKER 231
17 Lighting
    MICHAEL COWAN 241
18 Service tunnel and vehicles
    PETER SEMMENS AND YVES MACHEFERT-TASSIN 247

Section IV: Terminals

19 Terminal design and construction
    MARTIN STEARMAN 255
20 Terminal traffic management
    JOHN DAVIES 271

Section V: System operation

21 Operating the tunnel
    ALAIN BERTRAND AND ROGER HACQUART 291
22 Safety management
    RICHARD MORRIS 301
23 Conclusion
    COLIN J KIRKLAND 313

Contributors 317
Selected abbreviations 323
Index 324

Acknowledgements

The editor and publishers would like to thank the following for their contributions to *Engineering the Channel Tunnel*:

The individual chapter authors for finding the time to set down their experience for us.
John Noulton of Eurotunnel, Christopher Pick and Richard Hope for their help in reviewing the material and offering vital information and advice.
QA Photos for their photography and picture research, and all the other suppliers of illustrations.
Eurotunnel, whose *Channel Tunnel Trains* book (by Peter Semmens and Yves Machefert-Tassin) provided Chapters 13, 14 and 18.
Chapters 8, 21 and 22 are based on articles which first appeared in *Railway Gazette International*, May 1994, and are reproduced by permission.
Chapter 9, 'Power System' by JS Finn, was first published as 'The power system in the Channel Tunnel' in *Power Engineering Journal*, July 1991, and is reproduced by permission of the Institution of Electrical Engineers.
Chapter 17 is based on an article which first appeared in *Electrical Design*, May 1994, and is reproduced by permission.

Introduction

COLIN J KIRKLAND

A project on the scale of the Channel Tunnel, and capable of having so great an impact, is bound to capture the public imagination. It has already inspired a considerable number of publications.

So why should this book be of greater interest than the others?

Its intention is not simply to describe how the engineering problems raised by the construction of the Tunnel were overcome, but also to ask those who were responsible to explain the solutions considered, rejected and finally adopted. How this tremendous engineering achievement was accomplished is recorded in the context of the technical challenges, the timescale and the multitude of external influences that had to be taken into account. In the following pages, some of the engineers who took the key decisions aim to convey an idea of the host of options that had to be evaluated before the final solution emerged.

Before you proceed to discover the engineering complexities involved in the Channel Tunnel, the scene can be set by recalling some of the initial problems that had to be resolved before construction could even begin.

The idea of a tunnel from England to France is not new, and engineers have firmly believed for over 200 years that it is practicable. The first step on the road to proving feasibility was taken in 1881, when the two governments agreed to the construction of experimental tunnels. About 2000m of tunnel were excavated on each side of the Channel (that on the UK side known as the Beaumont tunnel), thus confirming the expected chalk stratum, and engineering practicability.

The second step was taken in 1973, when approval was given to a further trial tunnel. Sadly, this was abandoned by the governments in 1975 after about 250m of tunnel had been built at Dover, on the line of the present service tunnel.

The starting point for the successful Eurotunnel venture was the official 'Fixed Channel Link Report of the UK/French Study Group', published in 1982 by the UK and French governments. This determined that on economic grounds, a fixed link in the form of bored twin tunnels, with roll-on, roll-off traffic facilities, was most likely to be in the interests of the two countries. A closely following report, called 'Finance for a Fixed Channel Link' and prepared by a group of British and French banks, concluded that, subject to certain tests, such a project should prove financeable.

In 1984 the two governments were persuaded by the bankers that a privately funded scheme was possible. They therefore agreed to 'facilitate' the construction of a fixed link, and to grant 'Concessions' that permitted the building of the link and its operation for a period of 55 (now extended to 65) years. On the expiry of the Concession period the link would revert to the two governments.

There had been much debate among engineers, and in the press, over what sort of crossing should be built. Both the government of the day, and a considerable body of public opinion, believed that a 'drive through' motor tunnel was what was required – responding more to the prevailing demand for personal freedom than to common sense or safety. Just imagine a three-lane submerged highway, 50km long, with no laybys, and 12,000 individual drivers!

Ideas for a multispan suspension bridge, using special fibre ropes instead of steel for suspension, and enclosing the roadway in a perspex tube, were put forward. Another group considered a continuous barrage, topped by a road, with locks for ships to pass through and turbines to harness sea movement for power generation.

The problem of finding sources of finance was perhaps less in the public mind. However, extra taxation is always unpopular, and the idea of private financing for infrastructure projects was gaining ground in the context of a strong governmental privatisation drive.

THE BIDS

In April 1985 interested parties were invited to prepare their proposals and submit them at the end of October. The terms of the invitation were drawn as widely as possible, in order to encourage a range of bidders to come forward with innovative solutions. There were, of course, some restrictions. The link had to be fixed, not floating. The engineering had to be robust, using tried technology and methods in accordance with existing standards and norms, which was intended to protect investors in the scheme by keeping risks to a minimum. The project had to be totally privately financed, as there was to be no government support of any kind. The UK government would not permit itself to take any action that might be construed as giving any kind of financial guarantee to the concessionaire.

And so the first question was, what sort of fixed link should it be? A bridge, a tunnel, or a combination of the two? This was followed by: what sort of bridge, or what kind of tunnel?

Each of the four bidders set out their proposed solution, with a detailed justification of the engineering rationale, and a commitment to procure all the necessary funds from private sources.

Those six months in 1985 were characterised by frantic activity on the part of the engineers and bankers who prepared and submitted the Eurotunnel bid. It should be made clear that, although the Concession was awarded to Channel Tunnel Group/France Manche (later to become Eurotunnel), the bid was prepared by the five British and five French contractors[1] who would later become Transmanche-Link (TML) assisted by the five major banks who would later become known as the arranging banks.[2]

The contractors saw it as an opportunity to secure a large volume of construction work for their companies, and the banks an opportunity to arrange the commercial funding for the project. Neither group saw any urgent need to spend money on the creation of the concessionaire until they knew that the competition was won.

Although the prospect of creating a project that had been talked about for 200 years was exciting for all concerned, the risks for the bidders were enormous. The decision to create three parallel tunnels (two rail tunnels and a service tunnel) beneath the sea followed a lead set much earlier, in the late 1950s, when an early feasibility study suggested that this was the best solution, but how long it would take and how much it would cost were matters over which the contractors had to take the risk.

The form of the two transfer terminals at either end of the Tunnel was established as a continuous loop, rather than a reversing terminus, many years before 1985, but the detailed definition and quality standards to be employed were scarcely touched upon at the time when the cost estimates were made. Thus, in respect of the terminals, which were contracted for on a lump sum basis, the price risk was theoretically with the contractor, but with scope and specification so ill-defined that the possibility of arguments over perceived changes was very high.

Concerning the rolling stock and locomotives that might eventually traverse the Tunnel, all that could be said at the time of the bid was that the locomotives would conform to a readily available standard, and that the wagons would have to be able to accommodate wheeled vehicles from motorcycles to 44 tonne heavy goods lorries. Because of this lack of definition a sum of money

---

[1] *Translink Joint Venture (UK)*: Balfour Beatty Construction Ltd, Costain Civil Engineering Ltd, Tarmac Construction Ltd, Taylor Woodrow Construction Holdings Ltd, Wimpey Major Projects Ltd. *Transmanche Construction (France)*: Bouyges SA, Dumez SA, Société Auxiliaire d'Entreprises SA, Société Générale d'Entreprises SA, Spie Batignolles SA.
[2] Banque Indosuez, Banque Nationale de Paris, Crédit Lyonnais, Midland Bank, National Westminster Bank.

was included in the estimate which was little more than a guess, but which it was hoped would be enough to cover the provision of the necessary shuttle trains.

Which brought us to consideration of how long construction would take. Bearing in mind that the technical details of the proposed project were far from completely defined, some fairly imaginative estimates of cost and time had to be made, which were then increased judiciously by the contractor group to make allowances for uncertainty. There were considerable differences between the British and the French over the length of construction time required, with estimates varying between $7\,^1/_2$ and $8\,^1/_2$ years, the British being perhaps more confident of the tunnelling programme than their French counterparts,

The banks, of course were viewing this discussion from quite a different direction, and they kept a steady downward pressure on the cost estimates so that they could continue to accept the risk of procuring the financing for the project.

Just when we thought that these conflicting requirements were nicely balanced, the question of political risks was raised. In 1985 the two British and French governments were in agreement to facilitate the project, but what if one changed, or both, along with a change of philosophy? This led to a conviction that we should try to complete the project within the lifetime of two administrations, the logic being that a party might be re-elected once, but counting on twice was too big a risk. Thus seven years was to be the target, which was considerably less than the time arrived at by considering the construction demands. Reducing the time would be expensive, and so a balancing act resulted. How much did the contractor believe it would cost to accelerate the work by six months, bearing in mind that the earlier $7\,^1/_2$ year construction time was considered too short by half the team, and how big an increase in the funding requirement could be accepted by the banks without the risk of frightening off the investors and lenders?

The final prebid anxiety concerned the form of the contract under which the contractor group would carry out the design, construction, testing and commissioning of the whole of the project. Bear in mind that the ten contractors represented an extremely powerful group on one side of the 'negotiation', and that the 'employer' – Channel Tunnel Group/France Manche – existed only on paper, and you will appreciate the difficulty of the situation. The 'employer' was represented by myself, the only person in the team not drawn from the contractor group, and the Maître d'Oeuvre (the supervisory engineering body employed by Eurotunnel), struggling to be impartial. The only real pressure that the employer could bring to bear at this time was the unfinanceability of any contract which was not seen to share the risks fairly, or to control the flow of money properly.

The bid was submitted, on time, at the end of October 1985, and there followed the collective drawing of a huge sigh of relief. The feeling of relief was, however, short lived, when we realised that comparison with the other bids indicated that we had a good chance of success. There was no time for rest and relaxation. Margaret Thatcher had promised that the winner of the competition would be announced in mid-January 1986, and the 'clock would start ticking' on the seven-year construction period shortly afterwards.

The bids submitted for the Channel Tunnel were:

- The Channel Tunnel Group/France Manche (Eurotunnel) proposal was for two single-track rail tunnels plus a separate service tunnel, running from Folkestone to near Calais. A drive-on, drive-off vehicle shuttle service would be provided using specially designed shuttle trains, and through trains operated by national railways would also use the Tunnel. Construction cost: £2.6 billion (1985 prices).
- The Channel Expressway proposal was submitted by British Ferries Ltd (a subsidiary of Sea Containers Group) and involved twin bored road tunnels running from Folkestone to Calais, each tunnel with a two-lane carriageway. There would also be twin-bored single-track rail tunnels. Cost: £2.6 billion (1985 prices).
- The Eurobridge proposal (from the Eurobridge Studies Group) was for independent road and rail links. The rail link would be a single bored tunnel (plus a service tunnel), through which trains would operate in alternate directions. The road link would be a series of suspension bridges, supporting a total of 12 lanes on four decks, enclosed in a tube. Cost: £5.2 billion (1986 prices).
- The EuroRoute Group (British and French industrial concerns and banks) also had a road and rail proposal. The road would run in four lanes via a tunnel through the White Cliffs of Dover, across a bridge to an artificial island at the edge of the shipping lanes, then via a spiral into two lanes in an immersed tube on the seabed, to an artifical island and spiral, finishing with a bridge to arrive at Calais. The rail link would be of twin bored single-track tunnels (one including a service tunnel). Cost: £5.0 billion (1985 prices).

Channel Tunnel Group/France Manche won the bid, because their project, according to the two governments:

- offered the best prospect of attracting the necessary finance
- carried the fewest technical risks that might prevent it being completed (the other schemes were judged to be straining the technological boundaries)
- was the safest from the point of view of travellers
- presented no problems to Channel shipping

- was least vulnerable to sabotage and terrorism
- had limited and containable environmental impact.

The governments awarded the Concessions on 20th January 1986. On 12th February 1986 the Channel Tunnel Treaty between the UK and France was signed in Canterbury.

The bid had been submitted in the joint names of Channel Tunnel Group and France Manche, two national companies which were registered but had never traded. For all practical purposes they existed in name alone, but they were to receive the binational Concession to build and operate the Channel Tunnel system. The construction contractor group – which became familiar as TML – Transmanche-Link, having done all the work of preparation of the bid, was now gearing itself up for the enormous task that lay ahead.

For the embryo Eurotunnel the task was daunting. Within a matter of weeks a company had had to be created, capable of receiving this extraordinary Concession from two governments, and of entering into and controlling the biggest single construction contract ever – the design, construction, testing and commissioning of the Channel Tunnel – by now running at an estimated price of around £3.8 billion at outturn prices.

THE BILL

This new company had also to begin immediately the task of responding to the concerns of the many objectors to the Channel Tunnel Bill (4845 petitions in the House of Commons, and 1457 in the Lords), which had just begun its tortuous passage through the Select Committee process of both Houses of Parliament. The position was distinctly easier in France where all that was needed was a Presidential Déclaration d'Utilité Publique. The process took two days in France, compared with 15 months in the UK!

The task of responding to all questions raised during the passage of the parliamentary bill was carried out by a very small team of people, drawn from the contractor group (with the exception of myself) and representing the newly formed Eurotunnel. They sat in the public gallery in the House of Commons throughout the daily hearings, consulted with lawyers in the evenings, and spent the night preparing the formal responses, where possible, and making arrangements for the preparation of the more complex answers.

To give you some idea of how onerous this could become, the bid envisaged the disposal of the chalk excavated from the Tunnel at the point where it emerged, at the foot of Shakespeare Cliff just west of Dover. Remember that there would be nearly 5 million cubic metres of it – enough to fill Wembley Stadium 17 times – arriving at the surface at a peak rate of 2000 tons per hour,

24 hours per day, 365 days per year for three years. The Select Committee required Eurotunnel to investigate 70 possible disposal sites, spread across southern England, before it finally accepted the original solution (to fill artificially constructed lagoons at the Shakepeare Cliff site). Even after agreement was reached there developed a two-year debate on what shape and size the resulting chalk-filled platform should be in order to cause the least effect on Dover harbour.

There were 1001 other questions, and the pressure for a positive response was enormous. There was a very tight timetable within which the bill had to complete all its stages in both Houses of Parliament. If the timetable was not met, either the bill would fail entirely, in which case the project could not go ahead, or the passage of the bill might be delayed, which would seriously endanger the equally tight programme for the raising of the necessary financing. Many of the responses to petitioners would involve additional expense for Eurotunnel, and these extra costs had to be speedily evaluated and passed to the financing team for inclusion in their calculations.

THE FINANCE

Eurotunnel carried the responsibility for raising the essential finance for the project, a total of just over £6 billion to cover both construction and the establishment and development of the company itself. A small team travelled the globe speaking to bankers and investment analysts, persuading them of the capability of the engineering and construction teams to deliver a system which could be operated to achieve the required level of profit.

Once again, the pressure was enormous. We had to succeed at the first attempt: there was no possibility of a second chance. The timeframe was exceedingly short, constrained on all sides by the inescapable timing of other activities.

The contractor's seven-year contract period began in May 1986. Funds totalling £253 million were raised by a private placing in late 1986. These, together with the small initial shareholder fund, were enough to carry the company to about the end of 1987, when the major construction effort was programmed to start. The Channel Tunnel Bill had to pass into law before the major financing could begin. Loan funding totalling £5 billion had to be secured from a huge syndicate of international banks. Meanwhile, raising the £1 billion equity funding by public subscription could not begin before early November 1987 due to congestion in the Stock Exchange calendar. Eventually, the prospectus for the sale of Eurotunnel shares was the largest such document ever seen by the London Stock Exchange; its preparation was correspondingly complex.

While Eurotunnel was struggling with self-establishment, Parliament and fundraising, the contractor had begun the task of actual design and construction of the project. They naturally wished to have immediate access to all the work sites, and authority to go ahead with preparations for tunnelling, including the ordering of the machines that would bore the tunnels. Here, another major problem arose – how big should the tunnels be? What diameter? 7.3m or 7.6m? Since the proposal had referred to the two options it seemed a fair question, but the problem was that it could not be answered without considerable research, and the responsibility for the research lay with the contractor. In the end, safety reasons dictated 7.6m.

Very few engineers have had to concern themselves with the raising of money to pay for a project since the days of Brunel over 100 years ago. Here was a company that had no business until the construction of the project was complete. It had to rely absolutely on the reputation and ability of its engineers and forecasters to persuade others to invest in its idea. Its single greatest asset was the Concession, without which it could never have generated any interest.

Around 20% of funding was supplied by equity shareholders, loans the remaining 80%. Commitment to the loans was conditional on the equity being raised, and nobody was going to offer to invest unless the loans were secured. Another highly sensitive balancing act for Eurotunnel.

Because of the volume of funding required for a job of this size, the net was spread worldwide, which produced yet another problem for the fundraising team. To demonstrate that the conditions under which investors made their commitments were the same, regardless of geographical location, the team was obliged to undertake an almost nonstop, round the world tour to New York, Boston, Montreal, Toronto, Tokyo, within the space of five days, followed closely by presentations in the main European capitals.

It should be remembered that the finalisation of the funding exercise in October 1987 was fixed in time by the availability of a 'window' in Stock Exchange activity, and that quite by chance that window opened almost on the day that the Stock Market crashed. Keeping the attention of the investment houses focused on us was thus made doubly difficult. I shall never forget the day after 'Black Monday'. We were giving a presentation to representatives of US institutional investors in New York. I had done my usual presentation of the engineering challenges, using a slide projector in a darkened room. At the end of the presentation the lights went up and I observed that almost my entire audience were listening not to me but to their mobile phones recounting falling share prices! However, we simply had to continue – we would have had no second chance. Eurotunnel had no resources beyond those provided by an initial equity placement, which were practically exhausted, so the pressure for success was enormous.

PROJECT CONTROL

The principal concerns of the investors were in respect of the cost estimate, our ability to control costs, and the security of the eventual revenue which would repay their loans, and provide a dividend. The cost estimate had been prepared by the contractor group – TML – based on the description of the project contained in the proposal. The banks supporting the proposal of course wished to have a contractual structure that provided the greatest possible security for the cost estimate. In order to achieve this, the engineers devised a three-part contract that sought to share the cost risk between contractor and employer.

The Tunnel, which was perceived as carrying the greatest risk, would be built under a 'target cost' contract arrangement. In this case the employer agrees to pay the contractor the actual cost of work properly incurred, plus a proportionate fee, up to a target total agreed between employer and contractor. If the contractor is able to complete the work for a sum less than the target, the two parties share the saving equally, giving a strong incentive to the contractor to win a bonus for itself. If, however, the actual cost exceeds the target, then the employer pays only 70% of the excess cost, putting the contractor under considerable pressure to keep costs under control.

The construction of the two surface terminals together with all fixed installations throughout the system was priced as a lump sum. This meant that the costs of these works were at the contractor's risk, provided that there was no change to the scope of work or quality specification.

Finally, the cost of provision of locomotives and rolling stock was to be covered by a lump sum at the employer's risk, since the specification for these items was ill-defined at the time the proposal was lodged.

While this sharing of risk was entirely appropriate, particularly for a project on this scale, it brought in its train a huge cost control task.

Under the target cost form of contract the employer agrees to reimburse the contractor all properly incurred costs. There developed a complex set of arguments over what 'properly incurred' meant. Did it, for example, cover the cost of putting right a mistake? There were also protracted discussions regarding the mechanism for the agreement of any additions to the target price, an issue which was of considerable importance to our bankers.

We had to have a mutually agreed system between employer and contractor under which all expenditure related to the target cost was agreed before it was incurred – and before we could do that, the ten major contractors who constituted TML had to agree among themselves whose system they should adopt. Not surprisingly, each of these large international contractors had a computerised system of resource control and allocation, which was linked to its ordering and accounting system. Each naturally believed that its own system was the best, and so they mobilised their individual systems as soon as they began work on the project. Regrettably, it was almost $2^1/2$ years into the project before the contractors reached agreement, by which time the cost control arrangements were utterly confused, and expenditure was found to have run ahead of the target projections. In order to ensure that the incentives referred to earlier were maintained, the target value was renegotiated early in 1989, and the opportunity taken to resolve some early contractual claims at the same time.

As far as the terminal construction was concerned, the cost was of no concern to the employer provided that no changes were made. Our difficulty here was two-fold. There was frequently a lack of clarity in the description of the work due to the haste with which the bid had been prepared. When this situation arose the employer would request an amplification of the general description of the work. If the employer was not satisfied with the amplification and requested modifications, such modifications were often seen as changes by the contractor, changes that would require the employer to issue a variation order, which would normally attract an increase in the lump sum. The problem then was how to evaluate the appropriate change in the original lump sum, since the basis for the original lump sum was no better defined than the description of the work that created the problem.

The passage of the Channel Tunnel Bill through Parliament also brought about significant changes in the bid design. For example the entry into the Folkestone Terminal from the M20 was moved from the northwest corner to the southwest corner, and placed on a bridge instead of in a cutting, in order to restrict the environmental impact of noise to a single corridor; and the whole of the site drainage regime was reversed in order to protect a trout stream. Furthermore, we were subject to the supervision of the Intergovernmental Commission (set up to scrutinise the detailed design of the Tunnel), which could require modifications to be made.

All these complications in the mechanisms for cost control were magnified by the scale of the project and the very short construction timescale that we had imposed upon ourselves. That the monthly payment to the contractor frequently amounted to £90 million will give some idea of what this meant.

From the point of view of the lenders, it was important that, at any time during the construction, the value of the partly built project was worth the money spent, so a measure had to be devised to provide them reassurance in this respect. The task of devising such a measure, and of continuously updating it as the work progressed, was taken up by the Maître d'Oeuvre (Md'O), who, with the Intergovernmental Commission, provided the overall supervision of the works, for the protection of the lender and the public at large. This was particularly important because the construction was effectively forward-funded by Eurotunnel.

The Md'O was a form of engineering supervision, employed by Eurotunnel under the terms of the Concession, whose task was to monitor all activity on the project and to make quarterly reports to government and the banks on the progress of the work. Its particular responsibility was to ensure that the project was executed in accordance with the requirements of the Concession, with the binational technical standards, and with quality assurance standards established by the contractor. Although the Md'O worked for Eurotunnel, it was required to adopt an impartial stance between employer and contractor. At the commencement of the contract, the Md'O found it necessary to provide considerable technical support to the embryo Eurotunnel, and as time passed this necessity began to conflict with its independent role. In order to resolve this difficulty the three firms which comprised the Md'O formed two separate teams: a small Md'O team of about 40 persons continued to monitor performance and report, while a much larger team of engineers, numbering up to 400, was seconded into Eurotunnel to form the Project Implementation Division, which managed all the essential engineering for Eurotunnel, and managed the contract with TML.

The Intergovernmental Commission (IGC) was established by the two governments under the terms of the Treaty of Canterbury, and was paid for by Eurotunnel. The IGC, and its advisory arm, the Safety Authority, were responsible for supervising the project on behalf of the two governments. They were to ensure that matters of safety, security and environmental impact were properly incorporated, and that appropriate regulations were formulated and accepted for the construction and operation of the system.

Such a body was entirely unfamiliar to both Eurotunnel and TML from the outset, and indeed there seemed to be few precedents to guide the governments in the establishment of operating procedures. The French had some experience due to their many land borders already being crossed by transportation links. At the time when the bids were being formulated the bidders had very little information on the likely composition, role or powers of the IGC, so it was practically impossible for them to put a price on its likely intervention. In the bid document there was an attempt to set out the bidders' beliefs about what the IGC should do, but it would have been dangerous to try to dictate to the governments in this regard. In the event, when the IGC was established, it followed fairly closely the standard French practice, which is to require the client to submit 'avant projets'. Before any permanent work could begin, Eurotunnel was obliged to submit to the IGC an 'avant projet' giving in considerable detail proposals for the design, construction and operation of each major component of the system. The IGC studied these submissions, and responded within 15 days, either with comments or 'non-objection' (acceptance). Eurotunnel was entirely dependent on the contractor, TML, for the preparation of these 'avant projets', since it had been contracted for the 'design, construction, testing and commissioning' of the Channel Tunnel project.

Thus Eurotunnel found itself in a very difficult situation. It was responsible for providing information to an authority that could not say what it required, since it was not familiar with the project details, and it had to obtain the information from a contractor which was reluctant to be diverted from its task of fast-track construction to provide it. Moreover, the reference to safety in the ICG's remit could be extended, it was discovered, to cover anything at all, and this led to a very burdensome additional requirement of the engineering teams to research and prepare justification for all engineering proposals.

Just to give two examples of this, we can look at the design of the lining of the Tunnel. The designers knew that they must allow in the design for the possibility of an earth tremor: a remote possibility, but real, since tremors have been recorded on a line from Flanders to the Isle of Wight. Because predictions of the likely strength of a future tremor are very difficult to make, the designers made what they considered to be a reasonable guess. The Safety Authority checked historical records and confirmed that the last appreciable tremor occurred in 1531. They then required that the design should be capable of resisting a similar event, and required the designers to discover how strong the 1531 tremor was. A great deal of effort, and numerous meetings in England and France, were required before a compromise solution was found.

The second safety problem concerned the durability of the tunnel lining. Again, the design engineers were well aware of the danger that salt water presents to reinforced concrete, and considerable discussion had taken place between the British and French design teams to reconcile national differences of opinion on how best to provide the essential protection from corrosion. Despite all this effort, the Safety Authority insisted on a further comprehensive study of the risks and solutions by a specialist panel of international engineers and scientists, paid for by Eurotunnel, before they would register their 'non-objection' to the design proposals.

It is interesting to observe two further points here, firstly that the Safety Authority never gave approval to proposals, it registered 'non-objection' when its objections were satisfied, and secondly that much of the technical justification was being sought long after the start of construction. Some idea of the immensely detailed work carried out by the design teams is provided by succeeding chapters in this volume.

TUNNEL FACTS

It will be useful at this point to present a brief description of the Channel Tunnel and its operation, as a background to the material in later chapters.

INTRODUCTION 13

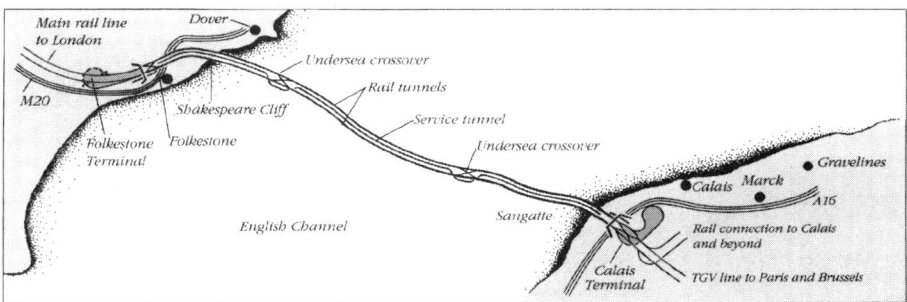

Figure 1. The Channel Tunnel. (Source: Eurotunnel)

The Tunnel runs between terminals at Folkestone (in Cheriton) and at Calais (in Coquelles), at an average depth of 45m below the seabed (Figure 1). It consists of three separate tunnels, each 50km long: two rail tunnels (also known as running tunnels) and a central service tunnel (see Figure 2). The two

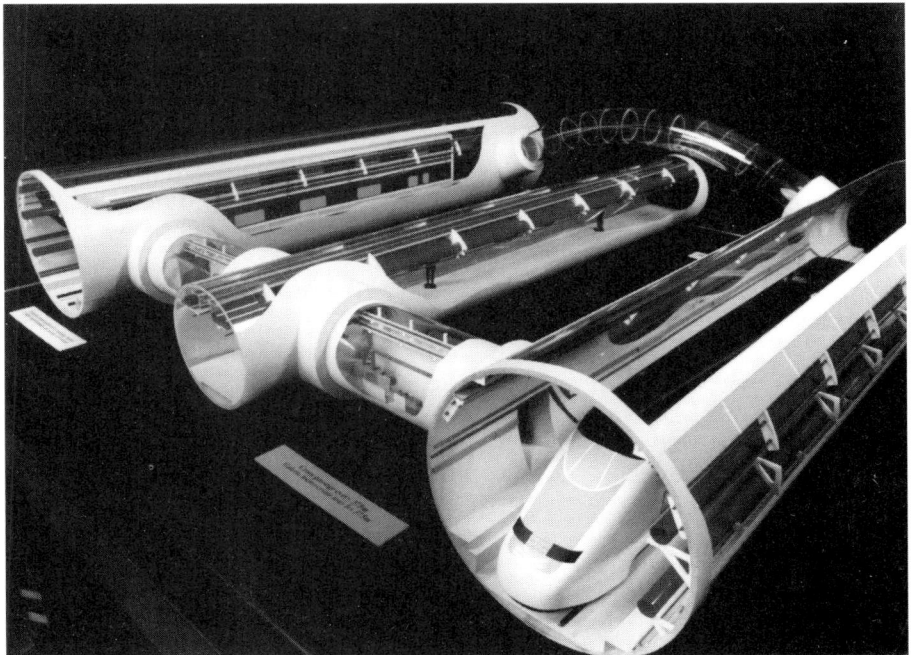

Figure 2. Model of the Channel Tunnel structure, showing the three tunnels, cross-passage and piston relief duct. (Source: QA Photos)

7.6m-diameter rail tunnels run 30m apart for most of the Tunnel's length, with the 4.8m-diameter service tunnel in between.

Every 375m, all three tunnels are connected by a cross-passage, used for ventilation and for maintenance and emergencies. The rail tunnels are joined every 250m by piston relief ducts (also called pressure relief ducts), which aid the aerodynamics of the Tunnel by allowing exchange of air. At two points in the Tunnel, there are crossovers, where the tracks of the two rail tunnels come together in a large chamber (or crossover cavern). Large steel doors that normally separate the tracks here can be opened so that trains can transfer from one tunnel to the other when one is closed for maintenance, or in an emergency. There are two additional crossovers, one near each portal.

Each rail tunnel takes trains in one direction only. Trains do not reverse but travel via turning loops at the terminals.

Eurotunnel runs an operation called Le Shuttle, which transports cars, vans, goods vehicles (and their passengers) and freight via specially constructed shuttle wagons. In addition, national railways run through trains (including the daytime Eurostar service) for passengers and freight.

STRUCTURE OF THE BOOK

Geology/survey

We look first at the geology the nature of the rock through which the Tunnel has been built. A wide variety of techniques has been used over the 200 years during which engineers have dreamed about a Channel Tunnel – only the geology itself has not changed. The original inhabitants of the microscopic shells deposited on the bed of the Channel millions of years ago could not have guessed that they would be used in 1987 to determine where the Tunnel should run!

The crucially important guidance of the tunnel boring machines is addressed next, requiring a very precise measurement of the route between England and France. Surface travellers can see where they are going and easily correct for errors. Beneath the surface the engineer is completely blind and must rely on great accuracy in calculation and care in establishing guidance systems. That the maximum 'error' at the junction of the tunnels was only 350mm is a tribute to the skills of the surveyors and engineers.

Tunnelling

Design of the lining for the Tunnel proved not to be an easy business. The Tunnel is not simply *a* tunnel, but many tunnels of many different sizes (rail tunnels north and south, service tunnel – all marine or land) constructed through a wide range of rock and water conditions. The basic philosophy may appear simple but there turned out to be many different ways to success and, because it is the Tunnel linking the UK to France, many opportunities to compare differences in approach between British and French engineers. There were protracted arguments over design methods, but it was recognised that the various acceptable ideas all led to the same result, which was both comforting and allowed agreement to be reached.

Following the decisions on the tunnel lining design, which enabled the engineers to fix the dimensions of the machines to bore the tunnels, the logistical problems of setting up access to the tunnelling works had to be carefully considered. How do you begin the construction of the longest undersea tunnel in the world? What will you do with the $7^3/_4$ million cubic metres of chalk which will result from the tunnelling? How will you install the 11 huge tunnel boring machines? The answers will be found in the chapter on initial access developments.

The major question, as far as the bankers were concerned, was our ability to construct the three tunnels between England and France within the three years allowed in the construction programme. Key to this achievement lay in the specially designed tunnel boring machines, and their ability to perform reliably. In order to meet the programme 11 machines were ordered, to specifications tailored to the expected ground conditions. The specifications, manufacture and eventual performance of these machines, which actually outperformed their targets and completed the tunnelling well within time, are set out here in considerable detail.

These machines, however, could not have performed their task if the logistical support had not been adequate. The spoil removal arrangements, and tunnel lining supply, together with power supply, lighting, ventilation, railway tracks, and a multitude of other services, had to be painstakingly planned, and interlocked to avoid failure.

Railway/tunnel services

The story then moves on to the complexities of designing the railway transportation system to be installed in the Tunnel. Electrical power, drawn simultaneously from the two national systems, is the lifeblood of the operation. It powers the railway locomotives, the lighting and ventilation, and the pumps which ensure that the Tunnel remains dry. It must also be appreciated that distribution of power throughout three tunnels, each 50km long, requires the provision of substations at intervals along the tunnels. The engineers designing the power supply systems had to decide early on in their work where these substations were to be located so that the tunnelling engineers could excavate rooms of adequate dimensions.

With power provided for the trains, the next problem is that of train control. This exercise begins with the forecast requirements of the train operators, both the two national railway companies and Eurotunnel, the operator of the continuous shuttle service between Folkestone and Calais. A wide range of train weights and speeds has to be accommodated, and demand varies widely through the year. The signalling system must be able to cope with complete safety with all conceivable traffic demands, and take account of the connections with the national railways, each of which operates a different system of signalling. Signalling no longer requires lights at the trackside: all information is displayed inside the driver's cab, using the running rails as part of the system of communication.

Communications – between controllers of the transportation system on each national shore, and from them to each train on the system as well as to the national emergency services – are crucial to safe operation. A combination of fibreoptic cable technology and independent radio systems provides the necessary capacity for transmission of voice and data messages.

Moving on to the more tangible aspects of the creation of the transport system, succeeding chapters present the broad philosophy and logic behind the design of the railway trackwork, in both the tunnels and on the surface at the two shuttle terminals, the locomotives and wagons comprising the shuttle system, and the other locomotives specially built for the Tunnel. It will be noted that considerable use was made of worldwide experience in these areas of design, and the relative merits of any alternatives were tested before final manufacturing decisions were taken.

The provision of a comfortable, breathable atmosphere in the Tunnel must be considered, including a cooling system and the control of air movement in the tunnels. Every aspect of air behaviour was studied, from basic requirements such as volume and temperature, to whether or not this long tube with holes in it at regular intervals would behave like a gigantic flute!

Apart from the risk of flooding, which is extremely remote, the greatest perceived hazard in the Tunnel is that of fire. Very extensive studies of all possible fire hazards were carried out, and a wide variety of detection and suppression systems were tested before the final decisions were taken. All the official fire safety agencies, and not a few unofficial pressure groups, took a very great interest in these development tests, and all had to be satisfied before authority to operate a public service could be obtained.

Lighting in the tunnels was expected to be required to be permanently switched on. Later studies related to safety and train driver reaction led to lighting being available throughout the tunnels, but only actually switched on for maintenance or evacuation.

The service tunnel, which runs alongside the two rail tunnels, has its own transportation system, provided for routine maintenance and service as well as emergency evacuation. The original concept of a rail-mounted system was changed during the detailed design phase to a more flexible wire-guided road system, with interchangeable body pods in the vehicles.

Terminals

The detailed design of the terminals at Folkestone and Calais, which had been seen as involving fairly straightforward civil engineering at the time of the bid, was discovered in the event to be much more complicated than expected, mainly because the complexities of such commercial operating issues such as toll collection and traffic control grew out of the development of Eurotunnel as an operator.

Finally came the preparation of the operating plans for this huge transport system, which again had to survive the scrutiny of the Safety Authority before permission to operate was given.

As the reader proceeds, learning of the vast range and complexity of the issues that confronted the Channel Tunnel engineers at every turn, they should remember that the project team had also to be constantly aware that their best efforts would be subject to the constant, unblinking scrutiny of two teams of supervisors, and the often ill-informed attention of the world's media.

Not that we should complain, for nobody would question that the safety of the future travellers on the system was always paramount, and the fact that all requirements are now satisfied is a tremendous tribute to the skill, care and patient application to detail of all who toiled to turn the Channel Tunnel proposal into reality.

I
SURVEY/GEOLOGY

2

Geology

COLIN WARREN AND PAUL VARLEY

The first proposals for a cross-channel link made in the early 19th century were too far ahead of the engineering techniques and geological knowledge of the time to be considered seriously. Many schemes were highly imaginative combinations of bored tunnels, immersed tubes, bridges and artificial islands; one proposal even involved the boring of a short direct route between Folkestone and Cap Gris Nez in Jurassic rocks of mixed lithologies, not ideally suited for tunnelling. Success on tunnelling beneath the Channel depended on gaining a sound knowledge of seabed topography and geology and, in particular, finding the most appropriate rock strata in which to build a tunnel.

Verstegan first noted the similarity between the chalk visible in the cliffs on either side of the Channel in 1628, and since then early researchers have confirmed this. It was not until Thomé de Gamond carried out his pioneer marine soundings and samplings during the period 1833–67 that the general seabed depth and continuity of the various geological strata across the Channel were established. Clear evidence was now available, based on the results of this survey and from a borehole subsequently drilled near St Margaret's Bay, immediately east of Dover, to indicate that:

- the Channel was relatively shallow: 55m at its deepest point
- the chalk of the cliffs on each side of the Channel was indeed continuous beneath the Straits of Dover and appeared to contain no major faulting
- the cliffs are made from four types of geological stratum of Cretaceous age laid down as marine sediments 90–100 million years ago. From top to bottom, the strata were the pervious upper and middle chalk, the slightly pervious lower chalk and finally the impermeable gault clay
- the 25–30m-thick chalk marl unit forming the lower third of the lower chalk appeared to provide the best tunnelling medium. At this level, the chalk has a clay content of 30–40%, which makes it relatively impermeable to groundwater yet easy to excavate and with sufficient strength to stand with minimal support.

Figure 1. 1881 Beaumont tunnel at Shakespeare Cliff, as seen in 1974 (Source: QA Photos)

Such findings subsequently led to companies being set up in Britain and France in the late 19th century with the aim of driving an undersea tunnel using Colonel Beaumont's compressed-air machine (Figure 1).

SITE INVESTIGATIONS

A summary of the boreholes drilled as part of the investigations for a Channel Tunnel is given in Table 1. A total of 116 marine and 70 land deep boreholes were been drilled along the alignment and over 4000 line kilometres of marine geophysical survey completed. This ignores the many hundreds of predominantly shallower holes that were drilled for the purpose of investigating and instrumenting areas of particular significance, for example the terminal, Castle Hill landslip and spoil disposal sites.

Table 1. Number of alignment boreholes sunk

Campaign	UK land	UK sea	French sea	French land
1958–62	7	5	3	1
1964–65	10	32	41	10
1972–74	8	9	7	–
1986–87 (phase 1)	19	3	9	15
1988 (phase 2)	–	5	2	–

GEOLOGY

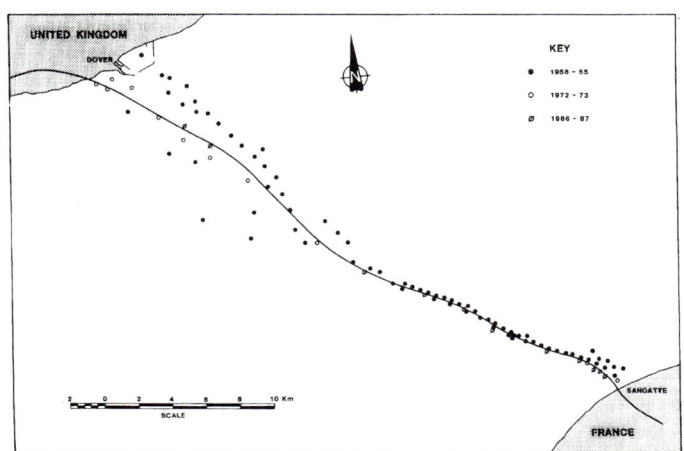

Figure 2. Marine borehole campaigns in the Channel (Source: Institution of Civil Engineers)

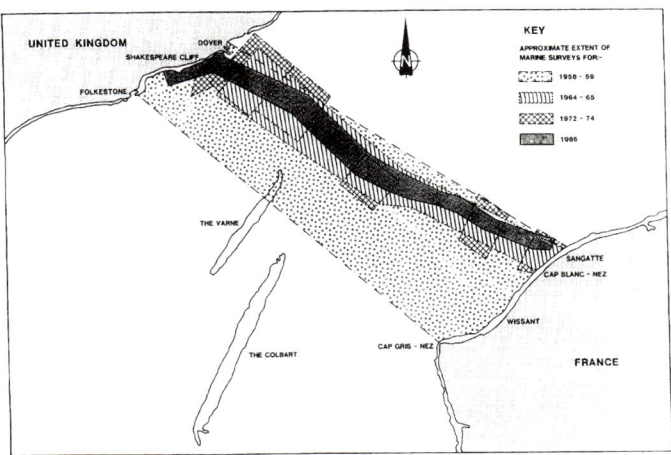

Figure 3. Marine geophysical surveys (Source: Institution of Civil Engineers)

Investigations were carried out during the periods 1958–59, 1964–65, 1972–74 and more recently in 1986–88 as part of this project. Figure 2 shows the location of the various marine boreholes as related to each drilling campaign to the end of 1987. Figure 3 summarises the marine geophysical surveys carried out.

The 1958–59 investigations considered not only a bored tunnel option but also an immersed tunnel and cross-channel bridge whereas most subsequent investigations have been aimed solely at providing a bored tunnel across the Channel. Consequently, a wider area of the Channel had to be investigated and hence both boreholes and geophysical traverse are widely distributed. At this time, the use of marine geophysics for engineering surveys was in its infancy and, although seismic profiling to depths of about 50m was achieved, both resolution and positioning were poor. The immersed-tube option was still being considered in 1962 when four boreholes were drilled on UK land.

The 1964–65 investigations concentrated on a much more northerly route which left the UK coast adjacent to Dover harbour. The work entailed several hundred kilometres of 'sparker' seismic reflection survey and over 70 boreholes yielding over 6000m of core. Micropalaeontological analysis was used for the first time to establish borehole correlations and aid identification of the several geophysical 'reflectors', enabling a good three-dimensional picture of the geology below the seabed to be made. The results of the sparker survey suggested the presence of a zone of 'deep weathering' immediately south of Dover harbour with relatively high permeability.

In view of the zone of deep weathering identified by the 1964–65 surveys, and because of operational and access constraints, a more southerly route was investigated on the UK side. The feasibility of this southern alignment and its extension to France was subsequently confirmed by boreholes and geophysical traverses completed in 1972–73. Details of the nature of ground to be tunnelled through were also provided from tunnelling work carried out on both sides of the Channel prior to cancellation of the project in 1975. At Sangatte in France, in 1973, an existing deep shaft had been re-established and a number of adits formed, and work commenced on the construction of a widened access adit (the *descenderie*). At Shakespeare Cliff in England there were comparable activities, including the construction of a widened adit and a large marshalling chamber for a tunnel boring machine (TBM). Although the project ceased prior to boring operations, the government did subsequently allow some 250m of 4.5m-diameter tunnel to be driven seawards through an instrumented section of ground to assess the behaviour of the chalk marl and confirm the overall viability of the scheme. The ultimate alignment, method of excavation and support on the UK side were essentially the same as those used successfully in this trial length.

The main objective of surveys carried out in the period 1986–88 was to improve, update and supplement the geological and geotechnical data existing along the route. In particular, emphasis was placed on defining accurately and quantitatively the level of the top of the gault clay and the properties of the host tunnelling medium, the chalk marl, found over 85% of the route. There were two separate phases of exploration:

- preconstruction investigations undertaken in 1986–87 enabling optimisation of the tunnel alignment against the suspected geological hazards and constraints (phase 1)
- detailed site-specific investigations completed in 1988 (phase 2).

For phase 1, numerous boreholes were drilled and instrumented adjacent to the landward drive including several water monitoring holes to check the effect of tunnel construction on the surrounding aquifer. Particular attention was paid to the old landslip area at Castle Hill in which the UK portal is sited, while further shallow boreholes were drilled immediately offshore at Shakespeare Cliff to verify the foundation materials of the proposed spoil lagoons.

In the Channel, investigations were formulated to provide continuous geophysical cover using both analogue and digital profiling over a 2km-wide corridor between the UK and French coasts. The seismic traverses were concentrated in a 1–2km strip with primary lines spaced at 1km centres on the UK sides, decreasing to 250m near the French coast. These ran across the alignment to define dip while the continuity of structure was provided by five longitudinal lines running parallel to the route. Accuracy to the top of the gault was believed to be within 5m. Twelve control boreholes were drilled, three in the UK sector.

To supplement this shallow geophysical survey work, three 10km-long digital traverses were completed. These enabled information on the geology and faulting of the basement rocks to be gathered at considerable depth and hence provided data on the likelihood of future earthquakes close to the Tunnel.

Phase 2 investigations were made at the proposed site of the UK and French crossovers, km 27 and km 44 respectively, and also mid-Channel, around km 38–41, representing a zone of high risk associated with the Fosse Dangaerd, a major undersea valley system. A total of seven boreholes were drilled through the chalk and gault clay into the lower greensand. Three boreholes were sited at the UK crossover, two at the French crossover and two adjacent to the Fosse Dangaerd including one within a minor tributary valley that crosses over the Tunnel at km 39.

As with phase 1, both analogue and digital geophysical techniques were used, although a far greater accuracy of prediction to within 1–2m was obtained by adopting closer traverse spacing. Spacing at the Fosse Dangaerd was generally below 200m, in contrast to 25m at both crossover sites, and hence both minor faulting and strata levels could be better defined.

A major improvement of the investigations was the adoption of marine geophysical techniques normally applied in the oil exploration industry, which, owing to developments in both electronic instrumentation and survey control, had advanced considerably since 1975. Such techniques include vertical seismic profiling, downhole geophysics allowing computation of synthetic seismic traces, and digitally acquired seismic data enabling both manipulation and use of geostatistical methods.

The offshore seismic surveys employed a high-energy source to derive a reliable and regular single-pulse signal, the multiple reflections of which were recorded by strings of geophones aligned in a streamer towed by the survey vessel. The data so produced could then be manipulated by computer stacking and enhancing techniques allowing data to be filtered to suppress unwanted noise, boosted to enhance positive reflection events, or as in the case of troublesome water bottom multiples, totally removed. The use of such techniques has ensured greater confidence in the results and enabled the information so produced to be reliably and accurately related to datum and co-ordinated to plan.

The 1986–88 boreholes were normally located on such geophysical traverses and were drilled down to the lower greensand below the gault clay and subjected to downhole geophysical logging. Use of the modern geophysical techniques enabled synthetic seismic traces to be produced from the sonic and gamma-density logs which could then be matched to the seismic events recorded on the geophysical profiles. For the first time, the more prominent geophysical reflectors could be identified within the stratigraphical sequence, thus enabling more precise geological modelling of the base of the glauconitic marl and better definition of the overall geological structure including areas of significant faulting and folding.

GEOLOGY

The basic geology on both sides of the Channel may simply be described as comprising north-easterly dipping Cretaceous strata, distinguished by the chalk and underlying gault clay. These rocks outcrop on the northern limb a large east–west anticlinal feature known as the Wealden–Boulonnais dome, in which both Cretaceous and Jurassic rocks are gently deformed over a structurally complex basement of Palaeozoic rocks (seen outcropping on the French side and in past workings in the Kent coalfield). A thin transitional sandy stratum known as the glauconitic marl (tourtia) separates the chalk marl, located towards the base of the chalk, from the gault clay.

Although the Tunnel lies entirely within these strata, its routeing and optimisation in both plan and section is a more complex problem than it first appears, the overriding principle being to:

- keep the Tunnel within the chalk marl, ideally the bottom 15m, known to be both continuous across the Channel and suitable for rapid excavation by machine

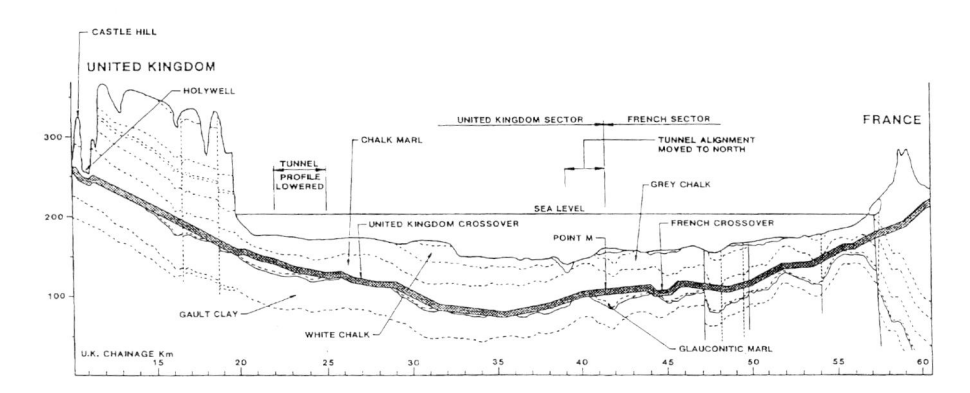

Figure 4. Geology along Tunnel as constructed

- maintain an adequate cover of sound rock above the tunnel crown
- avoid areas of weathered and faulted chalk or drift-filled pockets that could extend below the seabed down to tunnel level.

A reduction in cover to seabed and/or higher levels within the chalk marl would lead to increased water inflows entering the Tunnel as fractures and joints become more frequent and open. At lower levels, gault clay if encountered would increase the loads on the tunnel lining, would swell and soften in the presence of water, and give difficulty during tunnel excavation. Consequently, the primary objective was to fix the tunnel alignment as low as possible within the chalk marl (Figure 4).

On the English side, the strata are flat, lying with a dip usually less than 5° but towards the French coast this increases locally up to 20°. Past earth movements have led to fissuring, jointing and minor faulting of the rock mass on both sides of the Channel. On the UK side, faulting involving major displacement is generally absent from the area although minor faults with throws less than 2m are present; joint frequency within the chalk marl is 1–3 joints per metre. On the French side, faulting becomes more prevalent with throws up to 15m as the influence of the Quenocs anticlinal fold is approached. Most of the faults are of limited width and infilled with calcite, pyrite or remoulded clay.

The low dip of the strata on the UK side gives engineers increased freedom in the choice of route (most early schemes being sited further north under Dover harbour), in contrast to the geology on the French side where the increased dip, presence of faulting and reduction in chalk marl thickness gives engineers very little room for manoeuvre. Table 2 shows the geological succession on either side of the Channel.

Table 2. Geological formations existing across the Channel

Formation	Typical thickness (m) UK	France	Description
Upper chalk	–	90	White chalk with flints
Middle chalk	65	65–70	White marly chalk, some flints in upper part. Hard nodular Melbourn rock (11–15m thick) at its base
Lower chalk	75–80	65–70	Plenus marl at top: 1–3m thick White bed: yellow-grey soft chalk Grey chalk: grey marly chalk Chalk marl: grey chalky marls Glauconitic marl at base: 1–5m thick; marly sands and occ. sandstones
Gault	45	15	Dark grey-blue mudstone/clay
Lower greensand	–	–	Alternations of weakly cemented sand and clay

The chalk has also suffered the normal processes of erosion and weathering, especially during the last ice age when the sea level was lower resulting in subaerial exposure of the rock beneath the Channel. This caused both weathering and weakening of the chalk, though much of this lies above the Tunnel.

Although no major hazards were identified by the site investigations along the route, two major geological features formed during the Quaternary period gave cause for concern. One was the undersea valley, the Fosse Dangaerd, and the other was the Castle Hill landslip in which the UK portal is sited.

The Fosse Dangaerd had been identified by the 1964–65 geophysical surveys in mid-Channel and represented a deep, buried, infilled valley system extending 80m below seabed. The palaeovalley system, infilled with silts and sandy clay, is probably related to some deep structural feature and is believed to have developed as a result of fluvial erosion in the Quaternary. The main valley runs subparallel to the Tunnel, which was located 500m to the north. The 1986 geophysical survey showed a minor tributary channel, some 10m deep, extending northwards from the main valley, to cross the Tunnel at km 39. As a result, the tunnel route was maintained as far north and as low as possible to maximise cover at about 22m to rockhead.

The Castle Hill landslip, immediately adjacent to the UK terminal, is an ancient landslip comprising displaced and backward tipping blocks of lower chalk, glauconitic marl and gault debris with its basal failure plane within gault clay. Because of various constraints, the tunnel portal had to be located in the slip mass and consequently the landslip had to be stabilised by buttressing the toe with spoil and constructing drainage adits.

Following a review of both published information and past investigations, it became clear that no previous cross-channel definition was available for the host tunnelling medium, i.e. the chalk marl, termed *craie bleue* (blue chalk) on the French side. Early geologists on either side of the Channel simply adopted different classification schemes for the chalk marl and this led to confusion, particularly as to the top of the unit.

It was thus vital that a clear understanding of the chalk marl was available channelwide, in terms both of its biostratigraphy and its engineering properties. This was achieved by use of microfossil assemblages within the chalk, the most abundant of which are the foraminifera, calcareous shells 1mm in size. By studying these under the microscope, it was possible not only to divide the monotonous lower chalk strata into a series of well-defined biozones, but also to delineate the top of the chalk marl by identification of certain guide-species i.e. species that exist only over a short period of geological time and hence mark time surfaces within the rocks.

It became clear that the chalk marl on the French side, particularly adjacent to the coast, was much harder and brittle and with better-developed fracturing/faulting than the English equivalent, giving rise to higher permeability. This explains in part the different approaches to tunnelling adopted on either side of the Channel.

On the French side, because of the more difficult geology, with strata much higher in the geological succession also having been traversed underland and close to the coast, a tunnel boring machine (TBM) that was able to operate in both open and closed modes was selected. Tunnel linings were generally made of concrete, were bolted and gasketed and were erected inside a totally enclosed tailskin to form a completed ring. Such an approach minimised the impact of ground, in particular the potential for major inrush of water, since the TBM could operate even if water pressures up to 10bar were met, equivalent to a 100m column of water covering the machine.

The much simpler geology of broad folds with no major faulting on the UK side and the presence of the more clay-rich chalk marl of general low permeability led to the adoption of open-face TBMs and tunnel linings that could be built rapidly. Thus an articulated concrete lining which could be expanded against the ground as it emerged at the back of the TBM was chosen. This meant that no bolts had to be tightened, a 'wedge key' permitting closure of the 1.5m-wide ring.

GEOLOGICAL ASPECTS OF UK CONSTRUCTION

Excavation of the bored tunnels commenced in late 1987 and had been completed by the end of 1991. During construction differing methods were been employed on the UK side:

- **Machine-bored tunnels** excavated by six open-mode full-face TBMs, placing expanded concrete and cast-iron segmental linings. These

comprise all tunnels constructed from Shakespeare Cliff : marine service tunnel (MST), marine rail tunnels north (MRTN) and south (MRTS) and the corresponding land service tunnel (LST) and land rail tunnels north (LRTN) and south (LRTS).
- **Roadheader excavation** of tunnels and large caverns, excavated and supported by New Austrian Tunnelling Method (NATM). These include: UK crossover cavern, Shakespeare underground development, Castle Hill tunnels, portals at Sugarloaf Hill, shafts, temporary crossovers and pumping stations.
- **Hand excavation** using pneumatic tools and cast-iron segmental lining of cross-passages, piston relief ducts, pumping stations and other ancillary structures.
- **Raise bore technique** employed for construction of second ventilation shaft (initial 300mm-diameter hole, 110m deep, subsequently reamed to 1.1m prior to shaft excavation).

During progress of the works, regular geological exploration, instrumentation and monitoring were carried out. A computer database was established and extensive use was made of computer-generated maps, graphs and stereonets to aid prediction of ground conditions. At the height of production some 30 geotechnical staff were employed on the UK side with several thousand records being made of the tunnel face. Particular attention was paid to those areas of high risk, e.g. Fosse Dangaerd or where problems were anticipated.

Machine-bored tunnels

The service tunnels centrally placed between the rail tunnels are pilot tunnels for the main drives and were commenced at an early date to establish the geology along the route and to identify zones of broken ground and/or water-bearing fissures. The people who worked in the service tunnels were therefore the true pioneers, gathering information from the virgin ground and showing the way for others. Such information might be provided by the mapping of rock faces exposed during the drive or by way of exploratory probing carried out down in the tunnel, enabling designers, engineers and tunnellers alike to respond to the conditions actually met. Appropriate measures could then be taken well in advance of the main drives, for example:

- modification to the tunnel alignment
- optimisation of the tunnelling systems and methods
- change in lining type
- ground treatment of fractured rock and/or zones of high water inflow locally to improve mechanical properties and reduce permeability.

Ground treatment was carried out if driving conditions indicated that it was essential for the safety of the works. The method involved impregnating the ground with grout injected under pressure from suitably positioned holes drilled from the service tunnel. If large cavities or fissures were present a bentonite cement grout was normally adopted, while for finer fissures a sodium silicate chemical gel was used. Such treatment could be carried out in either one of two ways:

- ahead of the face using drills installed in the TBM shield of the service tunnels. This would only be undertaken if the forward probes had indicated particularly adverse ground preventing further advance of the pilot drive. Grout would initially be injected into the forward probes to form a watertight plug before allowing the TBM to continue.
- in holes drilled laterally from the pilot service tunnel once it had traversed a length of adverse wet ground. The advantage of this method was improvement of the ground surrounding the tunnel with minimal impact on progress of the drives. Lateral treatment was successfully employed to grout a 700m length of wet ground extending from km 22.7 to 23.4 (just off the English coast) ahead of the advancing marine rail tunnel TBMs.

EXPLORATORY PROBING

Three forms of probing were involved during progress of the service tunnel drives:

Extensive forward probing was carried out ahead of the tunnel face to check the geology and locate areas of faulting and broken ground and/or zones of high water ingress. The probes were normally 56mm in diameter and were destructively bored (i.e. by grinding the rock rather than removing solid 'cores'). The drilling parameters monitored included water outflow and salinity, penetration rate and the nature of drilling returns (e.g. when the probe entered the glauconitic marl the returns changed in colour from light grey to dark green). Samples of rock cuttings recovered from the drilling returns by sieving were then subjected to onsite microfossil analysis. This enabled both the stratigraphic level of the tunnel and the likely path of the probe to be determined.

Vertical downward probes were drilled 16m back from the TBM face, where the tunnel lining was erected, and provided 35mm-diameter cores down into the gault clay, because it was important to keep the tunnel out of the gault clay, where pressures on the tunnel lining could be excessive. Drilling was usually carried out at the same time as a forward probe to minimise TBM downtime.

Geological analysis of the cores was completed in the tunnel prior to despatch to the surface for detailed microfossil examination. This enabled comparison to be made between actual and anticipated geology as determined from previous borehole and geophysical survey evidence. For this comparison, the readily identifiable glauconitic marl marker bed was normally used.

Sideways probing (i.e. from the service tunnel in the path of the oncoming rail tunnel TBMs) was confined to those areas where difficult ground had already been noted in the service tunnel drive or where information was lacking. On the basis of information obtained from these probes a decision was taken as to the need for ground treatment. The probes normally extended above the crown of the adjacent rail tunnels and provided details on the nature of the rock by way of geological description of the 35mm-diameter cores recovered or measurement of water outflow, salinity and permeability.

These forms of exploratory probing (Figures 5 and 6) proved to be one of the major successes of the project, with the total length drilled on the UK side almost equalling the distance across the Channel.

UK UNDERSEA TUNNELS: ENCOUNTERED GEOLOGY
In all three undersea tunnels, ground conditions were initially favourable but quickly deteriorated from about km 20.4 and thereafter up to km 23.4 as the

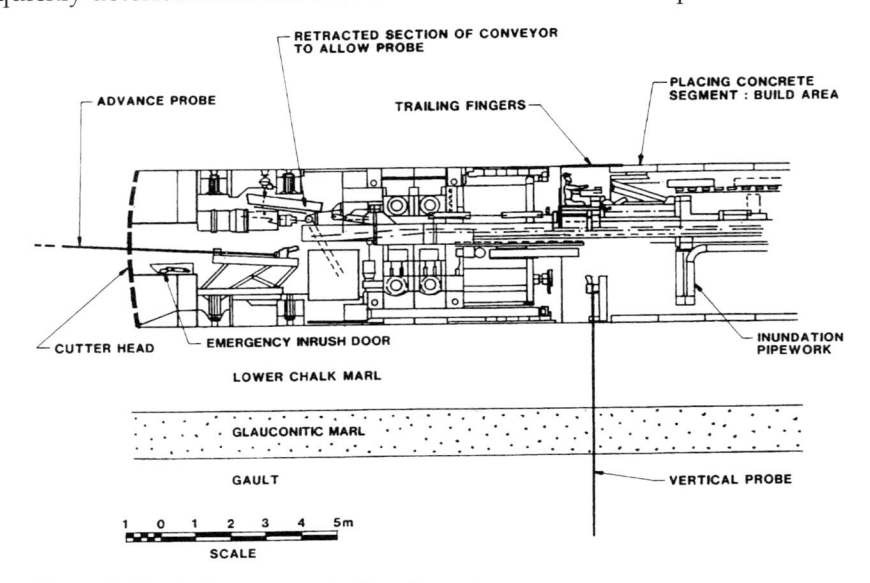

Figure 5. Typical arrangement of probing rig

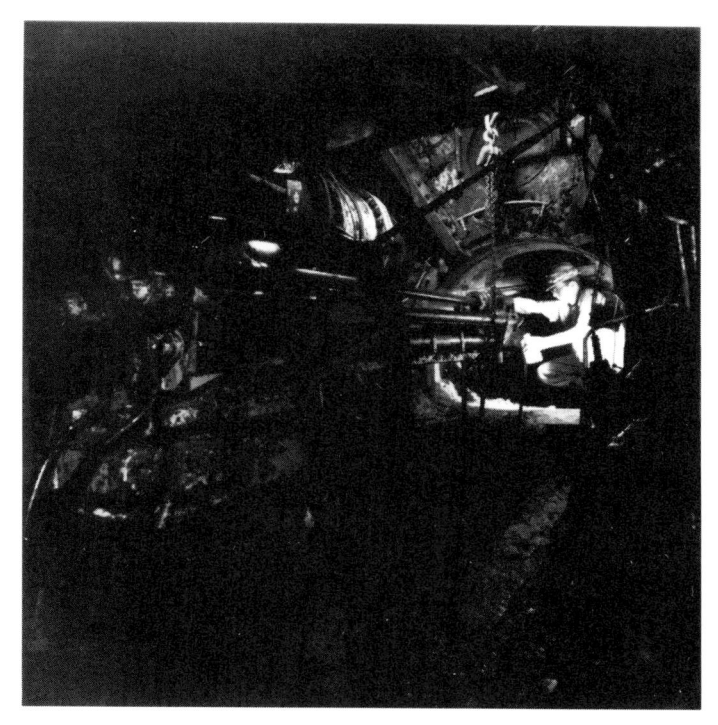

Figure 6. Probing: sampling flush returns for microfossil analysis. (Source: QA Photos)

cover to seabed reduced. The drive was continually hampered by overbreak and seepage. The porewater chloride levels of 4000–8000mg/l were quickly replaced by seawater, i.e. 19,000mg/l, resulting from downward percolation. The open joints also led to the development of high water pressures immediately adjacent to the tunnel walls, which together with the adverse nature of the jointing gave rise to block fallout. Overbreak and water inflow recorded in the MST construction towards km 30 are shown in Figure 7.

In the MST, wedge-type failures of the crown and shoulders were commonly observed, the shape of the blocks being defined by two joint sets, i.e. a subhorizontal bedding set and a steeply dipping set aligned subparallel to the tunnel wall. Block sizes were typically 0.2–0.5m^3 but when TBM progress was slowed larger volumes of overbreak would occur. This made expansion of the concrete ring difficult until trailing steel fingers (and later a complete hood) were fitted at the back of the TBM to restrain blocks from falling (Figure 8). Similar hood and finger arrangements were fitted to both marine running tunnel TBMs from the start of the drive.

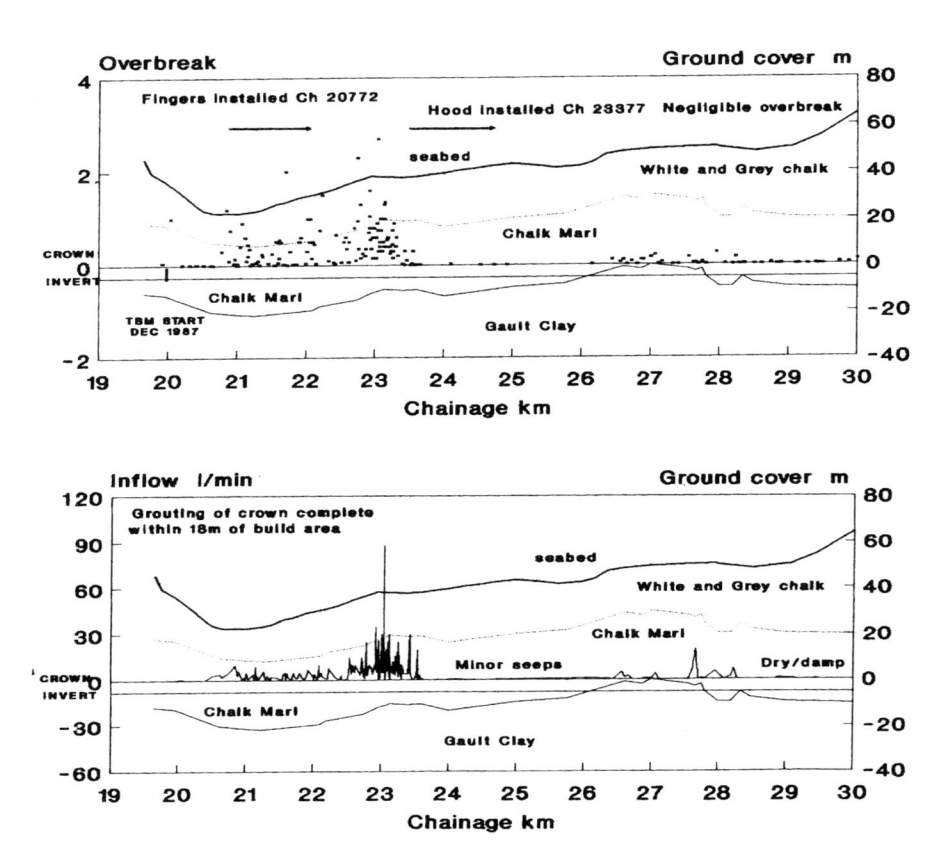

Figure 7. Water overbreak and inflow recorded in the build area of the marine service tunnel to km 30

There were two particular areas where the problems were aggravated by geological factors: a low-cover section (km 20.4–21.8) and an area of minor folding/faulting (km 22.7–23.4).

Along the early part of the drive, up to km 21.8, the vertical probes suggested that the MST was sitting much higher in the geological sequence than anticipated (i.e. usually 4m but up to 8m). As a result the tunnel was entirely located within the upper chalk marl with its crown lying close to the base of weathering and grey chalk. Consequently, more permeable and jointed strata were encountered leading to increased seepage inflow and

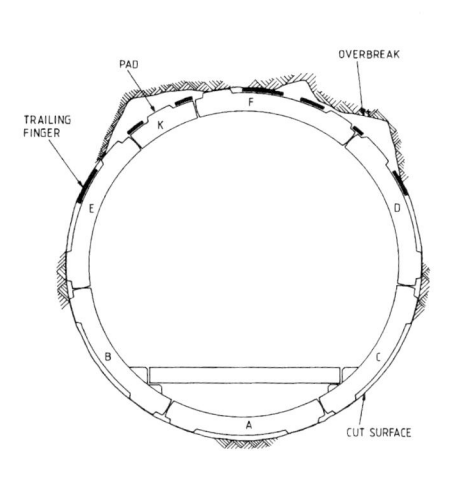

Figure 8a. Arrangement of trailing fingers, marine service tunnel

Figure 8b. Building ring within fingers. A: trailing fingers; B: tailskin. (Source: QA Photos)

overbreak. Immediate improvement was sought by steepening the tunnel gradient beyond km 21.8 thereby increasing the ground cover and lowering the tunnel into the more favourable lower chalk marl stratum.

As similar or worse conditions would be met in both MRTs up to km 21.8, lowering these tunnels below the level of the immediately adjacent MST was considered. This was rejected because of difficulties associated with building nonstandard cross-passages and piston relief ducts. Despite problems and slow progress, both MRTs successfully traversed this difficult ground. The increased size and raised elevation of the MRTs as compared with the MST brought about a tenfold increase in water inflow entering the tunnel (max. 120l/min) owing to increased permeabilies in the tunnel crown, i.e. from 10^{-7} to 10^{-6}m/s where cover to seabed was less than 25m.

Through the km 22.7–23.4 zone, water inflows entering the MST became greater as joints opened because of historical minor flexing of the chalk strata. Tunnelling conditions quickly deteriorated even though the MST was sited low down in the lower chalk marl (the preferred host medium). This section of ground, with permeabilities above 10^{-6}m/s, was eventually subjected to grout pretreatment prior to entry of both MRT bores. Such treatment involved the injection of a chemical grout, known as Silacsol T, into a series of arrays

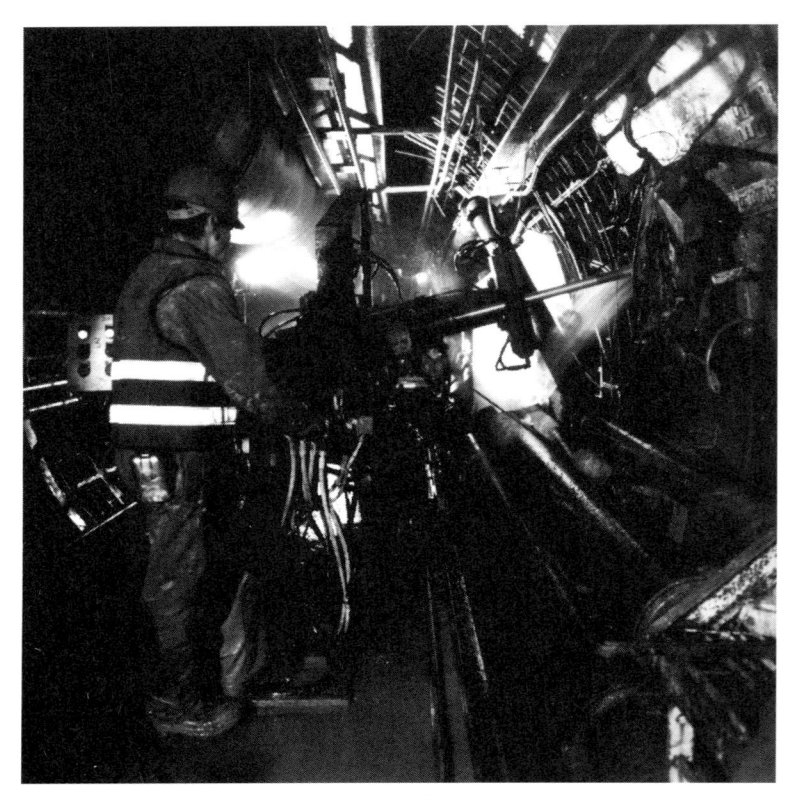

Figure 9. Ground treatment. (Source QA Photos)

of three inclined holes which had been drilled sideways from the MST across the axis/crown area of each MRT (Figure 9). Alternate precast concrete lining rings were drilled, i.e. 3m hole spacing, along a 700m tunnel length resulting in the treatment of 80,000m^3 of fissured chalk around a 3m-wide annulus.

Tube à manchette methods (Figure 10) using hydrofracture pressure up to 20bar were adopted to ensure full grout penetration into the surrounding rock mass. The *tube à manchette* technique allows for individual grouting of short 500mm lengths of drillholes, and for regrouting of any 500mm sleeve length if required. Little or no water was subsequently met when tunnelling through this zone, with both marine rail tunnel TBMs achieving better progress.

Beyond km 23.4 up to the end of the MST drive at km 41.6, tunnelling conditions were consistently good to very good. Damp conditions or only minor inflows and intermittent overbreak prevailed. The low salinity of the incoming water, equivalent to the chemistry of the porewater held in the

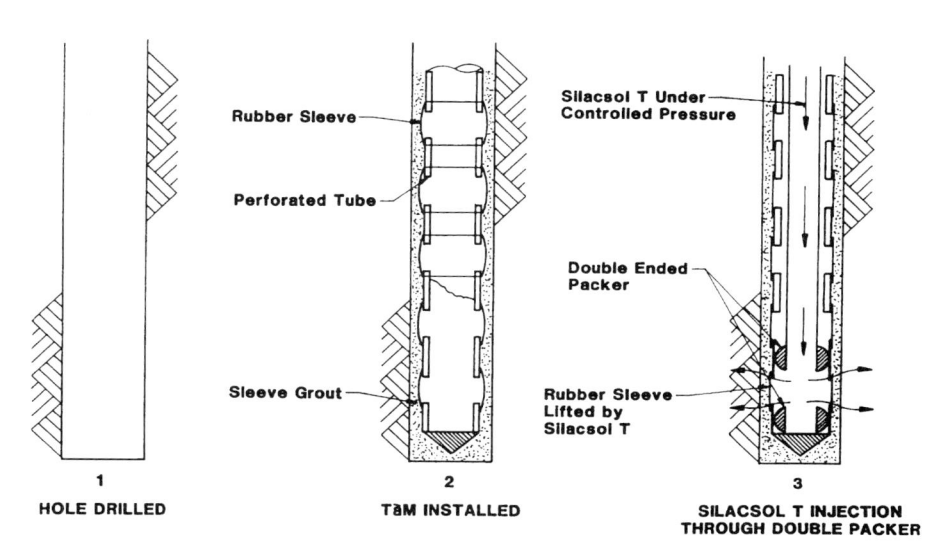

Figure 10. Tube à manchette injection system

rocks, i.e. 4000–8000mg/l chlorides, is a further indication of the low permeability and negligible downward drainage existing in this area.

Close to km 28, however, there did exist a further localised length of fractured ground where high saline inflow was met in the Tunnel. This can be explained by the presence of a geological structure, probably minor faulting/folding. Occasional problems were also experienced between km 34.5 and 37 as the MST proceeded through the glauconitic marl whose varying cementing and friable to strong nature created minor difficulty, i.e. sidewall spalling or stoppages for renewal of worn picks.

From this evidence it was clear that open jointing in the rock mass had caused much of the overbreak and increased water ingress. Such open joints were dominant where cover to seabed was less than 30–35m or in areas of pronounced faulting and/or folding as indicated at km 23 and 28. In view of this, the tunnel route adjacent to the Fosse Dangaerd was modified further to maximise the amount of cover above the tunnel crown. This was achieved by moving the alignment 90m further north, thereby increasing the minimum rock cover from 22 to 28m where the Tunnel passes beneath the buried valley at km 39. No major inflows of water were encountered throughout this length.

The historic breakthrough between the UK and French service tunnels occurred on 1st December 1990 at km 41.596, three years and 21,773 metres after the start of the MST drive from Shakespeare Cliff. The UK TBM was driven off to the side and the final connection was made by New Austrian

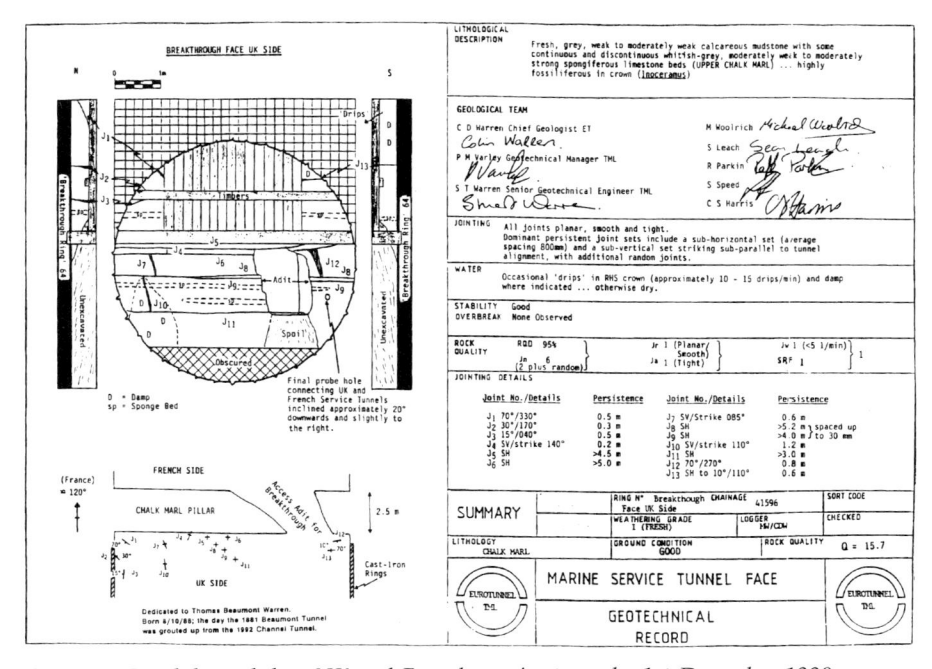

Figure 11. Breakthrough log, UK and French service tunnels, 1st December 1990

Tunnelling Method and simple hand excavation. The historic log of the breakthrough point is illustrated by Figure 11.

In May and June 1991, respectively, connection between UK and French rail tunnels occurred at km 37.219 for MRTN and km 38.500 for MRTS with the UK TBMs being driven steeply downwards to allow the French TBMs to be driven over them, making a full-face connection into the UK tunnels.

UK LAND TUNNELS: ENCOUNTERED GEOLOGY

From Shakespeare Cliff to km 13.7, ground conditions encountered by the land drives were good to excellent, i.e. generally dry chalk marl with only occasional incidents of block fallout. No evidence was observed to indicate the occurrence of the Lydden Fault, a major fault which supposedly cuts across the route adjacent to km 16.

Beyond km 13.7, all three tunnels encountered mixed-face geology as first glauconitic marl and then gault clay entered the face as a result of anticlinal folding centred at km 13.1 and the northeastward dip of the strata. Progress in all three tunnels was slowed by the blocky nature of the encountered strata with overbreak again occurring in the build area (i.e. the area behind the face where the lining segments were erected). This overbreak became more frequent as the tunnels approached the escarpment edge; joints opened up and minor water flows (less than 5l/min) entered the tunnels. Finger/hood arrangements had to be fitted to restrain block fallout in both LST (gault clay/ glauconitic marl) and LRTs (generally glauconitic marl/chalk marl).

Face collapse also became a regular occurrence as the minor water inflows lubricated 'greasybacks' within the gault clay leading to failure; large 0.4m^3-sized boulders of glauconitic sandstone constantly blocked the head and conveyor until broken down by pneumatic drills. Such problems were particularly pronounced where the gault lay above axis level (Figure 12).

Difficulties were also experienced during breakthrough of the LST into the TBM recovery shaft at km 11.1 where ground cover was minimal, i.e. less than 6m and the rock weathered and broken. This led to upward void migration above the tunnel crown as the tunnel face continued to collapse to such an extent that the ground had to be grouted prior to breakthrough.

To prevent the problem recurring during breakthrough of both land rail tunnel TBMs immediately adjacent, 20m-long entry chambers were constructed back into the main hillside using NATM. These extended back from the recovery shafts to a position where both cover and rock quality were considered adequate.

NATM excavations

The New Austrian Tunnelling Method is an empirical approach integrating the principles of the behaviour of rock masses and monitoring the behaviour

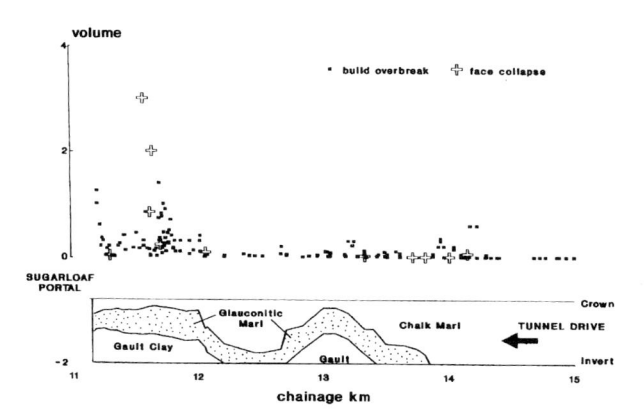

Figure 12. Land rail tunnel north: face collapse and build overbreak against encountered geology

of excavations during construction. Many people believe that, if shotcrete and rockbolts are used as support, they are employing the NATM approach. This is far from the truth. The method involves a combination of many established ways of excavation and tunnelling but the difference is the continual monitoring of the rock movement and the revision of support to obtain the most stable and economical lining. The major elements of the approach were first defined by Peck as the Observational Approach.

NATM was successfully applied on the UK side in three major areas: Shakespeare underground development (SUD), UK undersea crossover, and Castle Hill tunnels. The reasons for choosing NATM in these areas were manifold and were determined by the specific requirements. The principal aspects were:

- savings in construction time at the SUD, a complex network of tunnels built for temporary construction needs and permanent design requirements
- rapid construction of the UK crossover to avoid delay in the MRT drives
- adjustment of excavation procedures in the difficult ground conditions including landslip debris existing at Castle Hill.

SHAKESPEARE UNDERGROUND DEVELOPMENT

Ground conditions in this area were well known from the existing tunnels (1974 drive, Beaumont tunnel). The advantage of NATM was that not only could the complex series of interconnecting tunnels be more easily constructed but that major time savings could be achieved allowing early startup of the six TBMs, particularly as the primary shotcrete lining alone was capable of stabilising the ground sufficiently throughout the construction phase. NATM successfully completed:

- 2300m of tunnel including cross-connections
- 400m of adits above existing structures
- 60m of vertical shafts connecting existing openings
- 65 junctions and bifurcations.

One major area of concern was the construction of adit A2 whose portal is located on the lower Shakespeare Cliff site adjacent to the BR line. This tunnel had initially to be driven through uncompacted chalkfill containing numerous voids. Originally cut-and-cover excavation was planned. To reduce construction time, NATM was adopted for the 12.1m-wide opening after a trial shaft had confirmed that stabilisation by grout injection ahead of the tunnel face was economically feasible.

Roof settlements monitored by the instrumentation generally stabilised within the 100mm anticipated. In one isolated section where the invert sat on beach sand, additional ongoing settlement was recorded; the joint between the footing and the invert had been sheared off. The opening was stabilised by means of grout injections underneath the footing and at the sidewall.

Tunnelling downwards at 15% the adit soon entered the chalk marl rock where deformations and settlements became considerably less, in line with other NATM excavations in the SUD. Generally, settlements in these areas were less than 20mm.

UK CROSSOVER CAVERN

The UK undersea crossover is located 7.7km seaward of Shakespeare Cliff. At this location the MST, normally positioned between the two MRTs, deviates below and to the north to enable the rail tunnels to come in juxtaposition over a distance of 500m. The cavern itself is 165m long, 22m wide and 15m high, of a size to accommodate the height of three double-decker buses and the largest undersea cavern in soft rock built to date. During construction a number of temporary structures had to be built: MST enlargement to a length of 65m to accommodate plant and equipment, and 140m of construction adits with upward grade of 15%. The main cavern itself was constructed in stages by way of two sidewall drifts and crown heading/bench excavation.

The geology at the cavern site, determined by the borehole and geophysical investigation, indicated a normal geological succession dipping gently northwards and affected by minor faulting. The cavern is located within carbonate clayey mudstones forming the lowermost part of the cenomanian succession, namely lower and basal chalk, the more sandy glauconitic marl and a clay-dominant material 7m thick of carbonate mudstone at the top of the gault clay. The cavern crown is 35m below seabed, weathering penetrating down through the overlying grey chalk and into the upper chalk marl strata to within 20m of the roof. Formation mass permeabilities are generally low, i.e. 10^{-7} to 10^{-10}m/s, although higher permeabilities do exist in the glauconitic marl (owing to presence of open joints) and in the upper chalk marl immediately above the cavern crown. Exploratory probing from the service tunnel confirmed this.

No major problems were experienced during construction apart from one unexpected incident associated with settlement of the crown heading (see later). Generally, the rock mass quality encountered was fair to good with subhorizontal and subvertical joints spaced at one per metre and greater than two per metre, respectively (average persistence 2m). A number of minor faults, usually of an arcuate nature with downthrows less than 0.5m and striking 40–220° parallel to one of the major joint directions, were encountered during construction.

While reasonable quantities of water entered the adit and base of the sidewall excavations from discrete subvertical joints within the glauconitic marl, these significantly reduced with time, suggesting limited recharge from above. Further support for this was afforded by the general low salinity recorded, i.e. 20% that of seawater.

Water entering from discontinuities within the chalk marl at higher levels was limited to only minor seepages on joints some hours after excavation and thus excavations were reasonably dry. Exceptions to this were areas of the cavern crown where occasional water spouts with inflows measured at around 5l/min entered from joint intersections. The majority of inflow occurred via holes drilled 6m or more into the upper chalk marl for the installation of rock dowels and instruments. After excavation, total measured inflows from the cavern crown amounted to only 50l/min, the water generally approximating seawater salinity.

MST enlargement: While the glauconitic marl was harder to excavate, its competence overlying the weaker carbonate mudstone/gault clay material proved of significant benefit to stability during excavation. Minor problems were experienced owing to excessive floor heave associated with the swelling of the underlying gault clay. The heave totalled 50–100mm and had to be overcome by early invert closure and by the installation of long dowels. The upper surface of the glauconitic marl provided an excellent running surface for excavators and muck shifters during excavation of the lower branch of the access tunnel and the sidewall drifts of the main cavern.

Sidewall drifts: Typical deformations recorded during excavation amounted to only 10–15mm roof settlement and horizontal movements up to 10mm. Most deformations had stabilised after five days, that is about 15m behind the face, and did not increase with the passing of the second sidewall drift, indicating that the intervening pillar was strong enough to take the additional loads. Deformations during invert excavation, particularly diagonal deformations, increased sharply, e.g. an additional 20mm, suggesting high deformations occurring within the weak carbonate mudstone material. Such deformations eventually ceased once invert closure was completed with concreting of each 10m excavated opening.

Crown heading: During excavation deformations developed comparatively slowly, stabilising 25–30m behind the advancing face. Roof settlements amounted to 40–50mm. At the shoulders settlements amounted to 75% of this value while they reached only 20% above the footings.

At km 27.0, however, deformations suddenly and unexpectedly increased leading to cracking of the shotcrete layer 7m behind the face. The cracks developed over a length of 16m on both sides and about 1.5 and 3m above the footings. At this section, where the cracks appeared, the face was very dry and it became clear that high hydrostatic water pressures were being maintained 4–6m above the crown. This appears to have led to the opening of a horizontal bedding plane and increased loading acting on the lining which eventually caused the shotcrete to fail in compression.

The existence of a minor fault in the face at the time of the event appears to have been a contributory factor. Rock loads generally develop over a length of 25–30m when shotcrete strength amounts to 25–30MPa. Due to the fault, the full load was developed considerably earlier when the shotcrete strength was only 18MPa. Hence the fault acted as a 'release surface' for the horizontal rock beam above. The situation was stabilised by drilling weepholes upwards into the crown to relieve the high water pressures and by additional rock dowels and a second layer of shotcrete.

The completed crossover cavern is shown in Figure 13.

Castle Hill tunnels

Castle Hill, in which the UK portal is sited, forms an 80m-high chalk outlier to the main southern escarpment of the North Downs. The scarp slope has a regular 30° angle and is formed of lower chalk above the flatter exposure (3–5°) of the gault clay. East of Castle Hill lies a re-entrant valley called Holywell containing a Site of Special Scientific Interest and through which the Tunnel runs in cut-and-cover box section. To the west below the escarpment edge are the lowlands on which the terminal is sited.

A series of ancient landslips affecting lower chalk and gault strata can be found along the escarpment edge. All these slips are believed to be of the order of 8000–10,000 years old and date from the end of the last ice age when meltwaters cut down through the still-frozen ground, oversteepening slopes and allowing exceptionally high water tables to develop in the escarpment. As these conditions no longer apply, further slip development has not occurred and the present slipped masses appear to be in equilibrium with the pertaining environment.

One of the landslips covers the west side of Castle Hill and intersects the portal entrance to the three tunnels. Tunnelling through the slipped mass could not be avoided and consequently a decision was taken to construct the

Figure 13. Completed crossover cavern with sidewall shell. (Source: QA Photos)

500m-long tunnels using NATM. This option appeared to provide the best solution, given the variable and difficult nature of the ground to be tunnelled.

The general tunnel layout and geology through Castle Hill is shown by Figure 14. At the centre of the hill approximately 80m of lower chalk overlies a 1–2m thick bed of glauconitic marl which in turn rests on the overconsolidated weak mudstones of the gault clay. The gault clay in this area is 40m thick and overlies the lower greensand (Folkestone Beds).

The two rail tunnels were driven through Castle Hill from the east, following the gently dipping weathered chalk marl, glauconitic marl and gault clay interfaces. The service tunnel was at a lower elevation almost entirely within gault clay. The final 100m of all three tunnels pass through the disturbed material of the landslip comprising backtilted blocks of displaced lower chalk, glauconitic marl and gault clay.

The geotechnical properties of the ground were expected to vary throughout the hill, the variations principally relating to relief of stress during its formation, the weathered state of the rock, the amount of overburden existing and the disturbed nature of the landslip area. Only minimal water ingress was

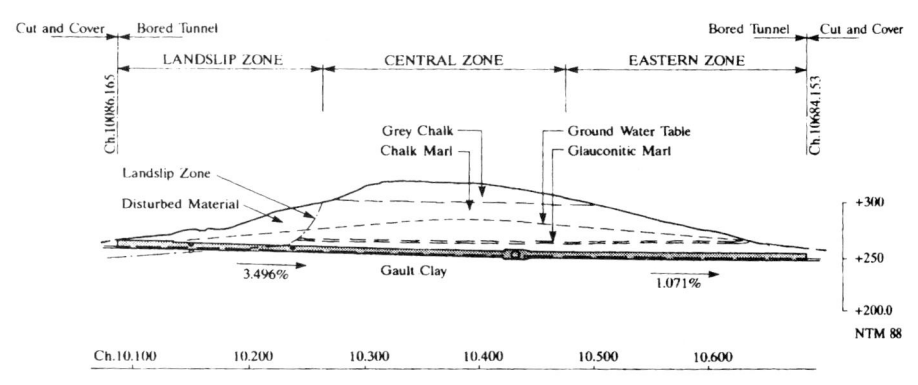

Figure 14. Castle Hill: service tunnel geology. (Source: IMM)

anticipated during tunnel construction although increased seepage was likely through the landslip material. For the purpose of design, the hill was divided into three zones, i.e. a landslip zone, a central zone and an eastern zone (Figure 14).

To augment the extensive site investigation and prior to selection of the tunnelling method, a 2–2.5m diameter trial heading was initially driven from a shaft sunk in the landslip zone on the line of the northern rail tunnel. The heading was lined with bolted segments, sections of which were instrumented and which included two zones built and instrumented by NATM, one in landslip material and the other in the undisturbed ground beyond it. The end of the last-mentioned section was left unlined, but the gault quickly showed signs of instability with joints opening and blocks falling out within days, and it was then supported with timbers for safety. The heading, however, proved both the nature of the landslip debris and the applicability of NATM.

It was recognised from the start that significant measures to stabilise the landslip would be required. The chosen philosophy was that: during construction, the stability of the landslip should not be reduced; on completion, the stability should be increased by about 10%; because of the difficulty in establishing the preconstruction stability of the landslip in terms of safety factors (no major recent movement was indicated), an observational approach would be adopted. Global observation of the slip behaviour was maintained throughout the construction period.

For stabilisation, approximately 50,000m^3 of toeweight material and extensive underdrainage from galleries driven off the main tunnels were proposed. The extent of the underdrainage to be developed would depend on water level observations. Material removed from within the 60m cut-and-cover box in Holywell Coombe was added to the toeweighting so that no passive resistance to slip movement was lost. Tension piles, predriven across the slip plane through the base of the box, similarly would serve to anchor the slip during the excavation/construction phase.

Because of the large volumes of toeweighting and tunnel excavation within the slip mass, short- and long-term creep movements were anticipated and compensated by increasing the tunnel size where it passed through the main slip surface.

The service tunnel drive started with full-face excavation through the gault clay, found to be jointed. At the face, joints were initially closed but became open owing to the relief of horizontal stress, minor water seepage sometimes being observed. Thus, outbreaks at the face occurred despite the use of a large number of 4m-long spiles installed in the crown. Following two major falls, one of 35m^3 and one of 8m^3, the method of support was changed to inclined support construction whereby the face was excavated with a 2m batter and the lattice arch installed with the same incline. This method proved highly successful and was adopted for the remainder of the drive and for the rail tunnels (Figure 15).

Driven rather than grouted rock dowels were generally used throughout the length of the drive, when it was found that such steel dowels if rammed into the gault and left for 24h (by which time the clay had expanded and gripped the ribbed bar) were able to withstand the design loads required. Grouted dowels also had to be used in the chalk landslip material.

A more favourable geology existed in the rail tunnels owing to the fact that chalk marl and glauconitic marl encountered in the roof were less jointed than the underlying gault clay which had proved problematical in the service tunnel. Due to the weak and weathered nature of the material existing at the start of the drive, heading and bench construction using a 1.5m bench was employed initially. Reducing the height of the face and supporting it by a dumpling allowed a safe drive, but settlements of the heading could not be stabilised even by additional bolting or enlarging the footings. Thus driving the heading was interrupted every 25m advance to stabilise the settlements by excavating the bench and installing the invert. It was anticipated that such time-consuming interruptions would become more frequent as overburden increased, requiring early invert closure, so the heading and bench method was discontinued after 50m and the inclined arch method introduced.

The experience gained during tunnelling works can be summarised as follows:

- Deformations were independent of the overburden except at the start of the drives.
- Horizontal deformations were higher than expected owing to high horizontal stresses.

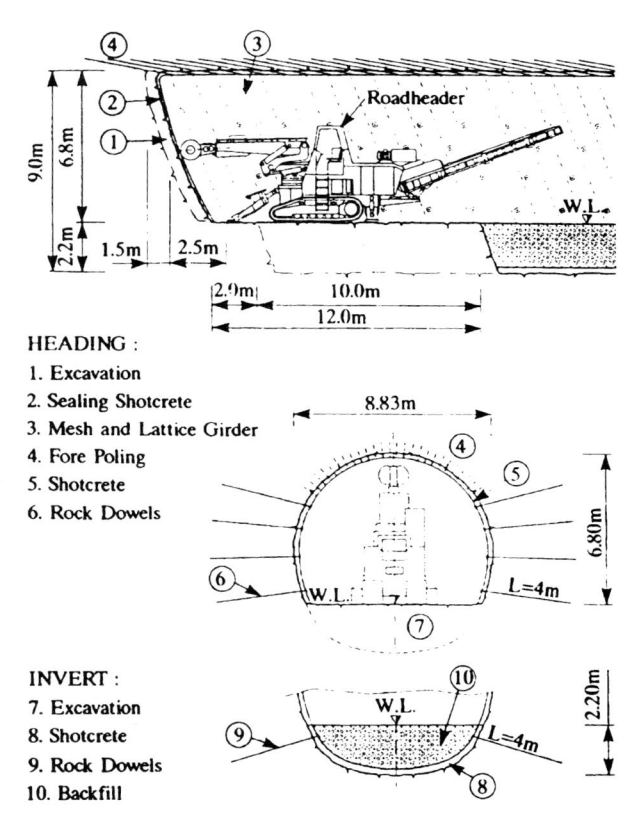

HEADING :
1. Excavation
2. Sealing Shotcrete
3. Mesh and Lattice Girder
4. Fore Poling
5. Shotcrete
6. Rock Dowels

INVERT :
7. Excavation
8. Shotcrete
9. Rock Dowels
10. Backfill

Figure 15. Castle Hill rail tunnel: excavation and support (Source: IMM)

- Tunnelling through the landslip was not as difficult as expected because of the generally intact nature of the slip debris and because high horizontal stresses had already been relieved.

Tunnelling through the landslip: Prior to commencing construction activity on the slip, extensive instrumentation was installed. This included inclinometers through the basal slip surface to measure elevation, magnitude and direction of movement, survey monuments to monitor movements at the ground surface, piezometers to measure pore pressure during construction and for long-term monitoring of drainage measures, and standpipes to monitor water-table variations.

As the toeweighting and cut-and-cover work proceeded it became evident that the construction activity had caused movement on the basal slip surface.

The movements were generally confined to that area being worked with some 90mm being measured in one inclinometer sited on the toeweight area immediately uphill of the toepoint. It was also apparent that such movement ceased or rapidly stabilised when construction work stopped.

An extensive search of old maps and documents did not reveal any evidence of movement of the landslip. However, during the construction of the works, when movements had been recorded and cracks developed in Castle Hill Road, a water main and gas main located in the road were exposed and inspected. The water main had been installed about 100 years earlier and the gas main 20 years earlier. Residual stress measurements on the gas main and the amount that joints had pulled on the water main were inconsistent with the measured movements during construction. This led to the conclusion that creep movement equating to 1mm per year had been taking place at least within the lifetime of the water main.

Each tunnel influenced the landslip as a separate event. In each case movement occurred on the slip plane as the tunnel face approached the underside of the slip boundary. Maximum downhill deflection was recorded when tunnelling was within two diameters of the slip plane. This 'shakedown movement' was confined to the upper central area of the slip as it moved down against the passive resistance provided by the toeweight-stabilised lower zone.

Movements recorded by the inclinometers during the passage of each tunnel totalled some 80mm at the back of the upper slip and reduced to 50mm close to Castle Hill Road. After the face had moved into the slip, further deformation was substantially confined to the area uphill of the tunnel reflecting normal ground loss associated with creation of a tunnel void.

The movement effectively ceased following completion of the tunnelling operations, although some minor downhill creep, amounting to 2–4mm per year, has been recorded. The works are monitored to check that the installed stabilisation measures continue to meet the design requirements.

GEOLOGICAL ASPECTS OF FRENCH CONSTRUCTION

In the French sector, the presence of faults with relatively large displacements, together with thinner chalk marl dipping at a steeper angle compared to the English sector, meant that there was a higher risk of encountering major inflows. The dictated the choice of closed-face TBMs that could permit exploratory drilling ahead of the tunnel face as and when required. Spoil extracted as a slurry via a screw was transported in muck wagons to the bottom of the Sangatte shaft where it was mixed with more water and and pumped to the Fond Pignon reservoir.

Geological monitoring during tunnelling involved the regular recording of data from a wide variety of activities. As with the UK sector, exploratory probe drilling from the service tunnel proved to be the most important technique for verifying the predictions of geological conditions prior to excavation. Such probes included occasional foreward probes drilled ahead of the TBM face, inclined holes drilled to the grey chalk, downholes drilled to the gault clay and other probes drilled to investigate faults and ground conditions at ancillary tunnel locations. Additional information was provided by face inspection, laboratory testing of water and rock samples and measurement of the rates of water inflow entering the face of each TBM.

No major geological problems were met, although water inflows were 10–20 times more severe than on the UK side, i.e. peak inflows of some 30–50l/s on the French side as compared with 2l/s on the UK side. This increase can be attributed to the Tunnel encountering fractured and faulted grey chalk and chalk marl. Major zones of inflow occurred underland and undersea from the start of drives to to km 56.7 and adjacent kms 55.6, 53.6, 49.2 and 47.5. As a consequence, all the underland and one in five cross-passages and piston relief ducts had to be systematically ground treated prior to excavation. Despite this, the ground conditions tunnelled through were better than those anticipated at the start of the TBM drives.

3

Control and construction surveys

ERIC RADCLIFFE

CROSS-CHANNEL CONNECTIONS

During the planning of the project prior to the 1973–75 construction, it was decided that the Universal Transverse Mercator (UTM) projection could provide a common horizontal framework for mapping and design more suitable than either of the existing, incompatible national grids. The UK Ordnance Survey (OS) and the French Institut Géographique National (IGN) developed mathematical transformations so that the UTM (Zone 31) grid could be added to existing maps and marine charts, allowing the locations of surveys, boreholes and alignments to be expressed in a single system.

Variable atmospheric refraction has always precluded accurate direct measurements connecting the British and French national levelling datums based on mean sea levels at Newlyn and Marseilles. However, a 1963 study of water movement in the Channel identified the existence of a sea slope of about 0.08m between the mean sea levels on either side. Combined with the tide gauge records and current national levelling data, this gave a difference between the level datums of England and France of 0.44m. For Channel Tunnel purposes a special reference level exactly 200m below Newlyn was specified to avoid the use of negative elevations where the Tunnel was deep beneath the sea.

The adoption of the UTM31 grid and the special reference level permitted the definition of preliminary horizontal and vertical tunnel alignments in a single system, but the inevitable distortions between the plane UTM grid and the spheroidal Earth would have led to complications in the setting out of alignments. A special projection was therefore created, fitting the Earth more closely in the Channel area, to minimise distortions. This was designated Channel Tunnel Grid (CTG), and defined as a Cylindrical Orthomorphic Transverse Mercator having a central meridian with a scale factor of unity at 1°30'E, and a latitude of origin at 49°N. Since scale factor and angular distortion in a transverse Mercator projection increase with distance east or west of the central meridian, the selection of 1°30'E, passing through the centre of the project, created a plane grid with negligible distortion for design and setting out purposes. The OS was commissioned to compute the new projection and produced CTG coordinates for a network of ten existing primary control points spanning the Channel (Figure 1).

These 1974 computations combined data from the British and French national triangulations with cross-channel observations made in 1951 and distance measurements from a European traverse of 1971. In 1986 the OS and IGN were commissioned by Transmanche-Link (TML) to update the Channel Tunnel Grid, re-evaluate the relationship between the national reference levels, and provide improved horizontal and vertical control on both sides. The basic geometry of the original CTG was retained, but the 1974 network by which it was manifested was considered less than adequate. A larger network was computed, incorporating existing secondary observations and other more recent data, including astronomic azimuths. To provide usable control points in the works areas some new surface stations were established to form a local network covering the Folkestone and Shakespeare Cliff sites (Figure 2), and the relevant observations were included with the previous data in a simultaneous least-squares readjustment. This computation provided revised plane rectangular coordinates in the CTG system, but to avoid confusion these were altered by adopting a shifted origin and this new network, now described as defining Channel Tunnel Grid 1986 (CTG86), was used for the initial design and construction.

Improvements in satellite location technology provided the opportunity for an independent verification and improvement of the network accuracy.

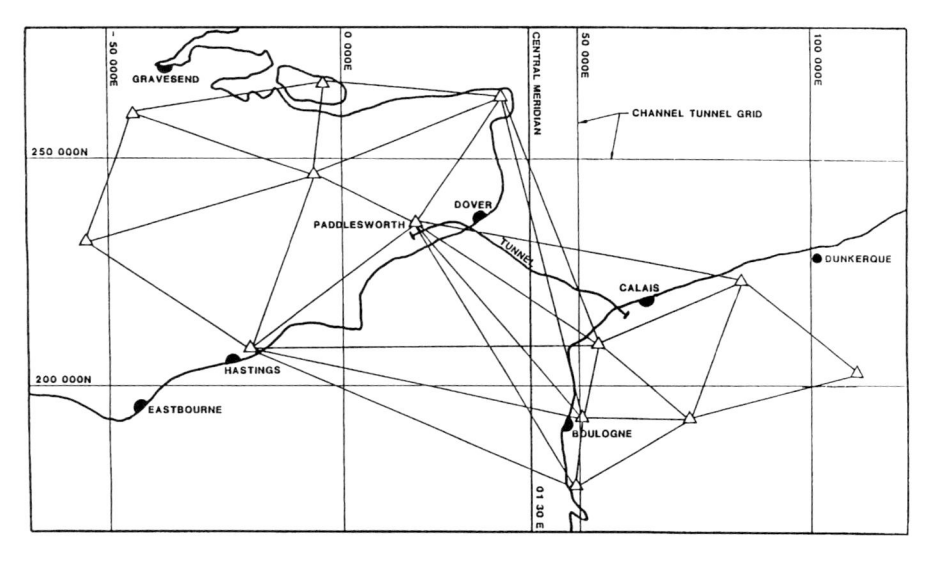

Figure 1. Main control network and grid. (Source: Eric Radcliffe/Thomas Telford)

The OS and IGN were commissioned in 1987 to carry out a campaign of Global Positioning System observations, the results to be combined with previous terrestrial data in a general readjustment. The adjusted network coordinates, though still on the CTG projection, were now referred to as defining the Réseau du Tunnel sous la Manche 1987 grid (RTM87). Accuracies of distance and direction over the network as a whole were estimated as 1 in 10^6 and 0.2 seconds, respectively, and this final RTM87 data was used for design and setting-out purposes for the remainder of the construction period.

Up to the commencement of construction in 1986, no direct measurement of the levelling datum relationship between the UK and France had proved possible, so the value based on sea-slope calculations together with up-dated mean sea levels at Dover and Dunkerque was adopted. The IGN69 datum was agreed as being 0.442m below Ordnance Datum Newlyn (ODN) (almost identical to the 1974 value) and the Channel Tunnel Height Datum (CTHD86) exactly 200m below ODN (199.558m below IGN69) was confirmed. On the UK side, the elevations of new benchmarks in the construction site areas were determined by a mixture of normal levelling and trigonometric heighting based on available OS geodetic benchmarks. The relationship between the national reference levels was slightly improved with data from the 1987 satellite observations, though some uncertainties remained owing to a lack of precise gravitational information in the Channel area. IGN69 was now agreed to be 0.3m lower than ODN, the tunnel datum being held at ODN minus 200m

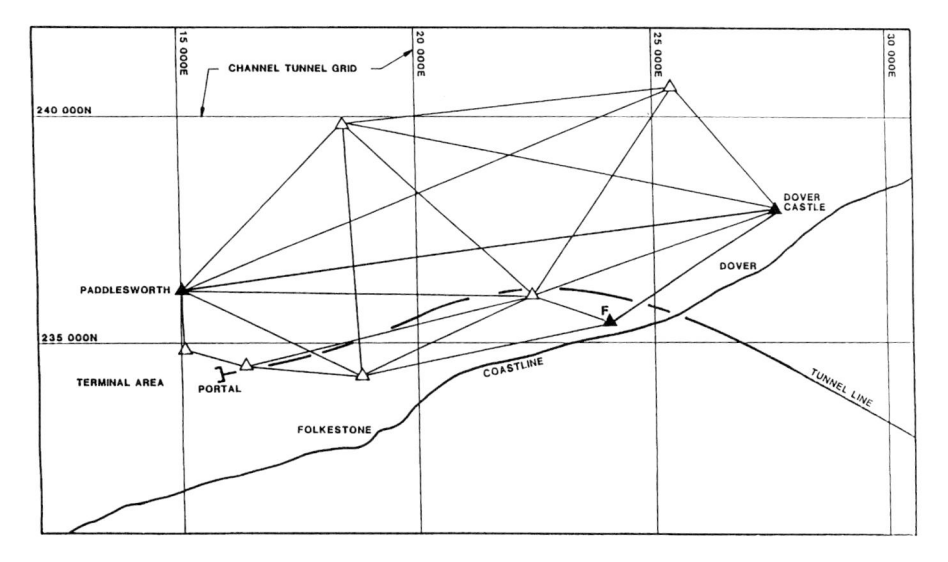

Figure 2. Local control network. (Source: Eric Radcliffe/Thomas Telford)

(199.7m below IGN69) and now designated Nivellement Trans-Manche 1988 (NTM88) datum. Elevations of French control stations and benchmarks were adjusted accordingly while those on the UK side required no further change.

TUNNEL CONSTRUCTION SURVEY

Since all the UK tunnels, except for a short length near the Terminal, were driven from the work area on the coast at Shakespeare Cliff, the transfer of survey control had to be made via the access available there. This comprised the two inclined adits A1 and A2 descending to the tunnel line from the lower work site, and the 100m vertical shaft sunk to tunnel level from the upper site. The adits were clearly preferable as a route for alignment transfer, though the congestion of construction installations on the lower site prevented the establishment of a long baseline at that level. Fortunately, one of the stations of the control network was close to the clifftop above the site and offered unobstructed sightlines from the portals of both adits (Figure 3). Since this station was visible over 4km from a station on Dover Castle, which itself was visible over 13km from one of the primary stations of the network, a very direct transfer of alignment from network to portals was possible.

The design of the adits themselves was not favourable to alignment transfer, owing mainly to a sharp change from steep gradient to horizontal near the bottom, which prohibited single sightlines from portals to tunnel line

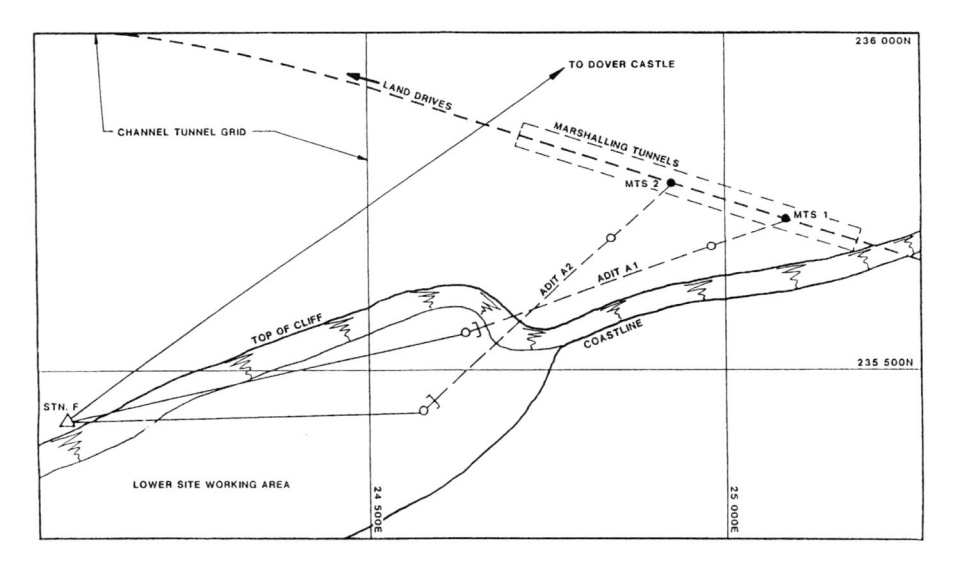

Figure 3. Transfer of control to tunnels. (Source: Eric Radcliffe/Thomas Telford)

and necessitated undesirably short traverse legs. Nevertheless, the configuration in plan permitted a full loop traverse through both adits, including a baseline in the marshalling tunnels from which underground traverses could extend to control both landward and seaward drives (Figure 3). Traverse stations consisted of stable concrete pillars with built-in centring devices, and observations were repeated a number of times under different atmospheric conditions to obtain an accurate location and azimuth for the underground baseline. Underground benchmarks were established by repeated precise levelling down both adits, and their values later confirmed by vertical electromagnetic distance measurements down the shaft from the upper site.

The maintenance of tunnel alignment comprised two distinct stages of survey: primary and secondary. The primary horizontal control was, in effect, an underground extension of the surface network, requiring survey stations of maximum stability with repeatable forced-centring capability for instruments. Brackets incorporating instrument 'tables' were mounted with expansion anchor bolts to preformed fixing holes in the concrete lining segments, the tables being specially designed to accommodate gyrotheodolites as well as the normal survey instruments and accessories (Figures 4 and 5).

The configuration of the primary network in the service tunnel was influenced by the layout of the various services required for construction.

Figure 4. Marine service tunnel looking towards France showing survey station bracket with target/reflector mounted. (Source: Eric Radcliffe)

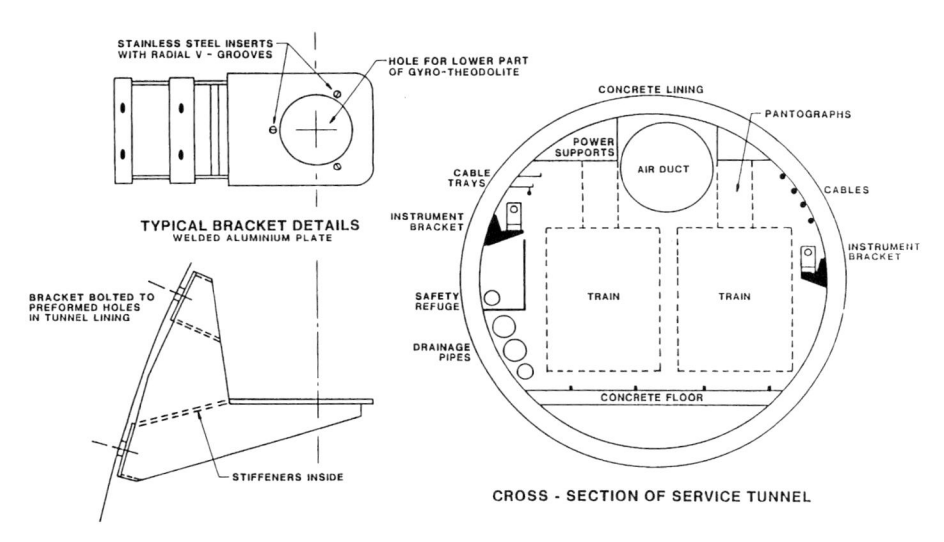

Figure 5. Survey stations in service tunnel. (Source: Eric Radcliffe/Thomas Telford)

Space had been allowed for personnel safety refuges on one side only, and initially it was decided that the control would be limited to a single traverse along that side, using the refuges as observation platforms. The lack of rigidity and vulnerability to the effects of lateral refraction made this configuration unsatisfactory for a long tunnel, but it was sufficient to allow driving to commence. Detailed study of the clearances available on the opposite side of the tunnel showed that it was just possible to accommodate stations that could remain clear of passing construction trains, and their installation permitted the observation of closed polygons and zigzag traverses which greatly improved network rigidity.

The results of these observations, combined with line azimuths measured directly with the newly delivered gyrotheodolite, showed that the sightlines of the single-side traverse had been curved consistently to one side owing to lateral refraction caused by temperature gradients close to the tunnel walls, whereas those of the zigzag traverse had a reverse-curve form. Thus the observed angles of the single-side traverse tended to be systematically too large, whereas those of the zigzag were alternately too large and too small by similar amounts, tending to be self-compensating in overall azimuth. The principle of zigzag traversing was therefore adopted for all primary control, and additional rigidity was achieved by overlapping single and double-length legs and including a mirror-image zigzag to provide symmetry. Close agreement between azimuths carried forward by this configuration and those

Figure 6. Laser beam passing along tunnel shoulder through TBM support train. (Source: QA Photos)

determined independently by gyrotheodolite demonstrated the value of the latter's ability to establish accurate azimuth irrespective of distance traversed; the frequent gyro azimuths were therefore incorporated as an integral part of the network rather than as independent checks.

The vertical control system for the tunnel drives was relatively straightforward, with permanent stainless-steel benchmark bolts being installed at 75m intervals and their elevations established by precise levelling using parallel-plate micrometer levels and invar staffs. Repeated observations under varying atmospheric conditions indicated that readings were not affected by any differential vertical refraction effects.

The secondary horizontal control consisted of a forward extension of the primary network to provide a reference for the steering of the tunnel boring machine (TBM). The congestion of the structures and equipment forming the TBM support train, about 250m long, limited this secondary traversing to a confined space along one side only, using temporary bracket stations. These provided mountings for the laser beam emitter (Figure 6) and its associated 'gates' which defined the reference line for the ZED guidance system incorporated into the TBM. The laser beam was set to a predetermined location and alignment so as to be received on a screen mounted on the cutting head.

Details of the beam position were loaded into an onboard computer which established the position and attitude of the cutting head in relation to the laser, and hence to the designed tunnel alignment data previously stored in its memory. A digital display of discrepancies between designed and actual TBM position permitted the driver to make continual alignment corrections by steering to 'zero'.

TUNNEL JUNCTIONING AND SURVEY CLOSURES

The first breakthrough and survey closure on the UK side occurred on the arrival of the land service tunnel TBM at Holywell after an 8km drive. To establish a preliminary horizontal closure a vertical borehole of 300mm diameter and about 30m deep was drilled from the surface to intersect the tunnel line about 200m before the breakout point. A precise optical plummet was set up above the borehole, its exact coordinates established, and a laser eyepiece attached. Once the TBM cutting head had passed across the borehole the vertical laser beam became visible in the lining erection area, and its position was surveyed in relation to tunnel drive control stations. This preliminary closure showed that no alignment correction was required, and the TBM subsequently broke out accurately into its dismantling chamber. The

Figure 7. Hundred-metre horizontal borehole being drilled from UK TBM to meet French drive. (Source: QA Photos)

CONTROL AND CONSTRUCTION SURVEYS 59

Figure 8. Drill rods from UK TBM arriving in face of French drive for completion of preliminary survey closure. (Source: QA Photos)

final control survey closure, carried out after the tunnelling equipment had been removed, gave discrepancies of only 4mm laterally and 15mm in elevation. The land rail tunnel control traverses were linked to the adjusted service tunnel control at the most forward available cross-passages prior to their breakouts, which were achieved with similar accuracies.

To establish a preliminary survey closure for the marine service tunnel the UK TBM halted at a prearranged point and a horizontal borehole was drilled forwards about 100m, while the French TBM was still driving towards the meeting place (Figure 7). The drill pipes were left in place to form a casing while the borehole was surveyed with a Reflex Maxibor downhole instrument to confirm that it had remained within the tunnel envelope, and to provide the French team with data on the position at which they could expect to find the hole when their TBM reached the end of its drive. The casing was withdrawn while the French TBM cut through the end of the borehole and replaced so that the French surveyors could make an accurate survey connection to their own control points (Figure 8). The comparison of the two sets of coordinates for a common point at the end of the borehole provided provisional data on the misclosure of drive alignments, and permitted the design of a preliminary corrective line for a further 50m drive by the UK TBM. While the French TBM was being dismantled from within its 'skin' the UK machine was turned aside on a tight curve to arrive at its prearranged burial position alongside the French one. The support train equipment, immediately to the rear of the UK machine, was then removed and the side of the diverted UK tunnel connected to the face of the French drive by a small hand-excavated adit, which was the scene of the official breakthrough on 1st December 1990.

On 3rd December 1990, the two main control survey networks were accurately joined through the small adit, each survey team carrying out the connection independently before comparing and agreeing results. The final agreed closure differences were:

- Chainage   75mm
- Lateral offset   350mm
- Elevation   60mm

The service tunnel alignment over the final 50m (excavated and lined by hand) was readjusted in the light of these results to produce a final smooth junctioning of the two drives.

For each of the marine rail tunnels the junctioning was achieved by diverting the UK TBM on a downward curve to be buried below floor level, and driving the French TBM through the temporary roof thus left to finish at the end of the UK permanent lining. As with the land rail tunnels, the control surveys were linked via the service tunnel at the most forward available cross-passages, and corrective alignments were applied where necessary over the relatively short drive distances remaining, both TBMs achieving accurate breakthroughs.

Throughout the tunnel-driving period as-built 'wriggle' surveys were made of the completed tunnel linings. These formed the basis for the final alignment design of the permanent rail tracks and other fixed equipment subsequently installed. Both wriggle surveys and setting out of the fixed installations were carried out from the drive alignment control stations after verification surveys to detect and eliminate any effects of local lining deformation.

II
TUNNELLING

4

Tunnel lining design

GUY LANCE

THE CHALLENGE

The provision of a fixed link between the UK and French shorelines represented a major engineering challenge for tunnelling engineers, and a challenge for which there is only one precedent. The 50km-long Seikan tunnel, which connects the Japanese islands of Honshu and Hokkaido, crosses the 23km-wide Tsugaru Straits with water depths up to 140m, and took 20 years to construct. The principal risk in constructing such long submarine tunnels is irruption at high pressure from the sea should unexpectedly bad local ground conditions be encountered. It is the duty of the tunnelling engineer to investigate and evaluate the likelihood of such risks and to ensure that they are controlled by the chosen design and construction method.

In addition to the physical challenge was the programming challenge dictated by the nature of the terms of reference for the project. The Concession to construct the tunnel, awarded to Eurotunnel by the British and French governments, was on the basis of private sector investment, thus necessitating the earliest possible return on capital which in turn dictated the need for the shortest possible construction programme. The future success of the project required that rates of tunnelling needed to be programmed that were at the forefront of current practice, yet applicable to the known characteristics of the ground to be traversed. This basic aspect of the tunnel design needed to be discussed and fully understood by the UK and French tunnelling teams in the short time available between the award of the Concession and the requirement to place early orders for the tunnel boring machines (TBMs), which had delivery times in excess of one year.

ENGINEERING BACKGROUND

All previous Channel Tunnel schemes had built upon the presumption that the presence of chalk cliffs at both Dover and Cap Gris Nez implied a continuity of the chalk horizon between the two coasts. Geological knowledge of the chalk strata and underlying deposits on both shores enabled prediction of the strata beneath the seabed between the coasts, which were tested by sampling (see Chapter 2). Thus, over the years, a large database of information vital to the tunnelling engineer had been built up. Not only had the geological strata been defined and sampled with an increasing degree of sophistication, but numerous pilot tunnels had been driven through these strata and were in most cases still available for inspection. In addition, man had been driving tunnels through chalk for many hundreds of years for mining purposes and engineers had developed a good working understanding of the methods appropriate for excavation and support. Nevertheless, it was generally true that these methods were based on empirical relationships and that a rational basis for predicting ground behaviour to the accuracy required by the needs of the project needed to be derived.

It is often said that tunnelling is more an art than a science, a view which acknowledges the practical requirements of the construction process and, in some cases, the extreme variability of the ground to be traversed. However, in the case of the Channel Tunnel the scale of the work, the economic necessity of ensuring that unit costs were minimised, and the need to satisfy the intergovernmental Safety Authority, meant that a predictive model had to be developed. Such a mathematical model required knowledge of the physical and engineering properties of the rock mass, the structure of the rock mass and the behaviour of water within the rock mass.

Obtaining this information was the purpose of the many site investigation surveys undertaken both before and during the span of the work. In addition, an understanding of the behaviour characteristics of the rock mass before, during and after excavation was necessary to predict the redistribution of ground stresses that would occur whether ground support was provided or not.

Experience pointed to the fact that unlined tunnels had been constructed within similar chalk material and remained standing for long periods of time, the Beaumont tunnel of 1881 being one such example. Observation of these existing tunnels confirmed that the generally applied principles of ground behaviour could be applied to the chalk mass once the local effects of rock structure and stress field orientation had been taken into account.

The final input required for the mathematical model was the calibration of predicted events against measurements of actual events, and this was able to be achieved through earlier research work undertaken on the 1973–75 scheme. In both the UK and France the opportunity had been taken to undertake measurement of both ground and support reactions in a controlled manner through the introduction of instrumentation into the works. The outputs from this test programme provided invaluable data for assessing the degree of accuracy for the analytical processes undertaken in 1985–86.

At the commencement of the Eurotunnel project, therefore, the tunnel designers had the two principal elements required for the provision of a 'ground model' on which to base a design: a large data bank of experience and a predictive analytical model. This ground model enabled the designer to determine both the optimum construction process for each element of the work and the most appropriate and efficient form of support for each element.

THE PROBLEM

The works to be constructed comprised the following basic elements:

- two 7.6m internal diameter rail tunnels spaced 30m apart, centreline-to-centreline, and each 50km in length
- one 4.8m internal diameter service tunnel sited centrally between the two rail tunnels, also 50km in length
- pairs of 3.3m internal diameter cross-passages, connecting the rail tunnels to the service tunnel, located at 375m spacing along the whole route
- piston relief ducts of 2m internal diameter to provide direct connection between the rail tunnels at 250m spacing
- two undersea crossover caverns to connect the separated rail tunnels
- three undersea pumping stations to provide for sumps and associated pumping machinery
- ventilation shafts sited close to each coastline
- access works at each coastline to enable construction of the permanent works.

CONSIDERATIONS

For each of these elements it was necessary to determine the form of lining that would meet the construction and physical constraints, and that met the overall aims of construction safety, low unit costs and maximum rates of advance.

Marine tunnels

In order to confirm ground conditions, during construction it was decided that the service tunnel would always be driven not less than 1km ahead of the trailing rail tunnels to act as a pilot. Construction of the three marine tunnels lay on the project critical path and therefore required high average rates of progress necessitating the use of mechanised TBMs. These machines are over 500m long overall so that they can accommodate the cutting head, propulsion gear, and all the necessary power packs and backup equipment, and this length led to two important design questions. First, would the ground remain standing in the period between excavation by the cutting head of the machine and the provision of support at a location a minimum distance of 20m behind the cutting head? And secondly, what type of temporary support, if any, would be appropriate?

Experience of mining the lower chalk led to the expectation that the ground would indeed stand up as required, but that support should be applied as soon as possible to ensure protection to the TBM backup systems. Construction logistics demanded that the support should be of the 'one-pass' type to avoid secondary activities having to be undertaken within the tunnel while the works trains were serving the TBMs during their long drives, and also that precast materials should be used as much as possible to assist materials handling underground, control materials quality, and provide full immediate ground support. This last requirement is particularly important for TBMs where the backup equipment requires a stable floor on which to run. A precast segmental lining was therefore chosen for the lining systems behind the TBM drives in both the UK and France.

Cross-passage and piston relief duct construction presented a different set of aims to the designer. Here, the work consisted of a large number of individual drives of short length which all had to be undertaken within a relatively cramped space in locations that did not encourage the handling of large boring equipment. An additional cause of restrictions was the need to access and service each separate workshop directly from the main tunnels whilst these themselves were being driven and serviced by the construction railway.

Following studies of alternative construction techniques the designers decided to specify a support of grouted cast-iron segmental linings for building by traditional hand-mining techniques. Each segment was sized to enable it to be manhandled into place during erection, a task which was made easier by the adoption of modern lightweight spheroidal graphite-iron segments. Again, a knowledge of the behavioural properties of the chalk assisted in the final choice of support and the mining gangs were able to determine the level of temporary ground support required for each individual excavation.

Details of cross-passage linings were based on traditional ribbed skins with connection by bolts through flat joints, whereas the piston relief duct linings required a smooth internal surface in order to minimise aerodynamic disturbance and this necessitated novel detailing. A form of 'inside-out lining' was devised which presented a smooth inside surface by dispensing with bolts and utilising tongue and groove joints between segments. This lining therefore had no inherent stiffness and it was found necessary to provide a temporary internal former during construction to maintain a proper circular shape prior to the lining being permanently grouted into the ground.

Access areas

An important requirement for the success of the project was to ensure that the TBM drives could commence as soon as practicable after access to the sites, and that sufficient underground space could be made available to provide for all the logistical backup equipment for the TBMs. At both the coasts of France and UK, provision of these marshalling areas proved difficult (see also Chapter 5). In France, the tunnels were located at a relatively shallow depth below ground level, but were sited in chalk strata which from the investigation work undertaken in 1974 was known to be highly fissured and water bearing. Fortunately, surface access presented few problems, so a solution was adopted to provide a large diameter shaft 55m in diameter and 70m deep through which all three tunnels passed and which provided direct vertical access down to all three tunnels. Water ingress during construction of the shaft was controlled by the provision of a grout curtain around the periphery of the shaft. On completion of the tunnelling, the shaft was used to provide normal and supplementary ventilation systems to the permanent works.

On the UK side, the marshalling area was sited on the line of the three main tunnels but 140m beneath the top of the cliffs. An access adit to the existing short length of service tunnel driven in 1974 from the working site at the base of Shakespeare Cliff was still available and experience of its construction indicated that the chalk in the area was dense, homogeneous and with low permeability. A new and larger adit was also required to serve the marshalling area, where sections of rail tunnel and service tunnel needed to be excavated to provide chambers for both the erection of the six TBMs and provision for their backup equipment.

Traditionally large tunnels in rock had been supported in the temporary condition by means of steel colliery arches providing support via wooden blocks to the excavated ground. The space between the arches was then lagged with timber planks to provide protection against loose blocks of chalk becoming detached and falling out. However, as a result of techniques subsequently established during the construction of the Alpine tunnels in Europe, an alternative form of construction had become available for primary lining. This technique has come to be known as the New Austrian Tunnelling Method (NATM; and see Chapter 2) and was founded on a set of principles which link the ongoing monitoring of ground and support behaviour with the ability to provide a range of support stiffnesses and thus enable control of the interaction between ground and support. It offered a construction method with the ability to provide an economical lining that could match the rates of progress required, particularly in respect of the large tunnel intersections which can provide tunnel engineers with immense analytical problems.

NATM was adopted for the construction of the marshalling area known as the Shakespeare underground development, all of which was temporary and thus required only a temporary lining. It subsequently provided a large amount of data on the behaviour of excavations in chalk and enabled back analysis to be undertaken on the accuracy of the original support dimensioning calculations, and thereby reassign values to some of the earlier engineering properties chosen for the chalk.

A form of NATM was also used for the construction of a 160m-deep shaft from the top of Shakespeare Cliff to the underground development. This shaft of 10m diameter was lined with a permanent support of mass concrete constructed by slip forming, and was used throughout the remainder of the project construction period for personnel access to the works via the hoist. This shaft was originally intended to form part of the normal ventilation system. However, because of programme slippage, a separate shaft had to be provided for this purpose.

The three major undersea pumping stations required the provision of sumps and adjacent facilities for pumps and their associated pipes and electrical switchgear. As with the cross-passages and piston relief ducts, access to the workshops and working arrangements off the main mechanised tunnel drives represented the principal construction constraints. Again, hand tunnelling methods for the construction of the necessary tunnels and shafts were deemed the most appropriate and the works were designed for a mixture of cast-iron segmental linings, NATM and direct *in situ* concrete lining.

One problem specific to the UK side was the topographical conditions close to the portal site at Cheriton. The bored land tunnels between Dover and Folkestone had to terminate at Sugar Loaf Hill where the railway crossed Holywell Coombe at shallow depth before re-entering bored tunnels to penetrate Castle Hill and exit via the main portal which was sited within an existing zone of landslip material. Tunnelling through the Coombe, part of which had been designated a Site of Special Scientific Interest, necessitated the adoption of cut-and-cover techniques to meet the technical environmental and economic constraints. Cut-and-cover works entailed the open excavation of the ground, the construction of a reinforced concrete box structure and finally backfilling around and over the completed structure.

The design decisions on the appropriate form of construction for Holywell Coombe were reasonably straightforward, but this was not the case for the Castle Hill tunnels where programme constraints, construction and plant availability and site-access restrictions led to a full evaluation of TBM against NATM alternatives. Successful use of the NATM for temporary support of the excavations at Shakespeare underground development, and the greater flexibility of the method in dealing with some anticipated difficult ground conditions, particularly through the landslip zone, resulted in the method being adopted for the Castle Hill tunnels (see also Chapter 2).

Crossover caverns

Perhaps the most difficult and complex problem faced by the designers was to determine the layout and construction procedure for the two undersea crossover caverns. The requirements of tunnelling engineers, railway operators and mechanical engineers were combined to specify a single cavern enclosing a diamond track crossing, across which large longitudinal doors could be drawn to provide compartmentalisation of the rail tunnels when the crossing was not in use. Such a structure required a single-span cavern 22m wide, 15m high and 165m long to be constructed ahead of the rail tunnels on the UK side, and subsequent to the rail tunnels on the French side. These two different sets of construction constraints led to two very different construction solutions for the caverns, each of which fulfilled the same functional requirements.

The French cavern arrangement was based on a method previously used on the Mount Baker Ridge freeway tunnel in Washington DC, and consisted of driving a series of small-diameter tunnels parallel to and around the periphery of the final cavern. Each tunnel is backfilled with concrete before an adjacent tunnel is driven alongside. On completion of all tunnels, a ring of concrete tubes has been formed and the ground can be removed from within. This method was used to encircle the completed rail tunnels before they were removed, together with the encircled ground, to provide the cavern.

On the UK side of the Channel the cavern needed to be constructed closer to the coast than on the French side and therefore came earlier in the overall programme. To safeguard the progression of the two main drives it became necessary to construct the cavern from the service tunnel, which was driven ahead of the two rail tunnels, and to complete it in time for the following rail tunnel TBMs to be able to pass through without being delayed. As this window of opportunity was small it was impractical to complete the cavern with a permanent lining within this timescale and a two-stage approach was required. This comprised an initial temporary lining which would enable passage of the rail tunnel TBMs, followed by installation of the permanent lining.

Excavation of such a large cavern beneath the seabed from long lines of logistical support was expected to be a potentially hazardous operation requiring great skill in planning and execution, and it was therefore decided that use of NATM, with its inherent flexibility of approach, was the ideal excavation method. It was possible thus to undertake excavation of the cavern in a series of pilot headings, enabling a limitation on the area of ground exposed at any one time to be imposed, and for pilot tunnels to be advanced to test the ground for areas of high water inflows (see Chapter 2).

Analytical models were derived which used the ground behaviour characteristics derived for the designs of the other tunnels together with the values for ground properties derived from the site-specific ground investigations previously undertaken. These calculations were used to predict the ground and lining loadings produced by the proposed construction sequences. The predictions were then closely monitored during the construction process and the degree of support amended as necessary to keep within the required loadings and movements. The tunnels were later provided with a heavy, *in situ*, permanent concrete lining.

LINING TYPES ADOPTED

Main TBM drives

For the main TBM drives, precast segmental linings offered the twin benefits of early ground support and ease of erection, but radically different solutions were adopted by the UK and French designers. The reason for this was that the ground permeability to water was known to be greater at the French coast, where the chalk was fractured and faulted, than at the UK coast, and this condition extended at least 4km from the French shoreline, but not as far as the mid-Channel point. The French tunnellers were therefore faced with the prospect of starting their drives in ground that was wet but expected to get dryer, while the UK tunnellers could expect to start and continue in dry consistent ground for the majority of their drives.

FRENCH SIDE

The French TBMs were of the earth pressure balance type which were capable of working under an external water pressure of 6 atmospheres, but could resist 10 in the stationary condition. Seals between the cutter head and the main body of the machine isolated the ground at the face while the cut chalk was ejected through a hydraulic piston discharger or double archimedean screw conveyor. For this type of TBM a lining is required which can be built within the protection of a tailskin located behind the main body of the machine, and capable of being connected directly onto the end of the tunnel already constructed. A tail seal needs to be provided between the inside of the tailskin and the outside of the tunnel lining to prevent ingress into the TBM of water under high external pressure.

In addition, the linings need to be made watertight against the high external pressures. This was achieved by placing neoprene seals around each segment which come into compressive contact during ring erection. To support the ring during erection, to connect it to the adjacent ring, and to maintain a compression in the joint seals during construction, bolts are located in all the joints. Once each ring is erected the TBM is advanced by using its rear hydraulic rams to push against the completed tunnel, leaving a void between excavated ground and the outside of the tunnel. This annular void is filled with a quick-setting grout to ensure full contact between ground and support to ensure a uniform distribution of loading on the lining rings.

These construction requirements dictated the segmental arrangement of the lining, but not the material to be used in its manufacture. Traditionally, linings were manufactured using cast iron to provide strength against external loadings, but cost considerations have led to an increased usage of reinforced concrete linings. These linings use very high strength concrete (>50MPa) to minimise thickness and enhance durability.

UK SIDE

The linings on the UK side needed to comply with a different set of criteria arising from the anticipated good ground conditions. It was not necessary to use an earth pressure balance TBM as the rate of water ingress was low and the ground could therefore be exposed. A primary requirement was for a lining that could be erected quickly to achieve the high rates of progress that had been programmed. A segmental precast lining was specified, but one which did not require any bolting and which could be erected behind the TBM without the need for a tailskin.

The lining concept was based on similar systems developed for use in the London clay where the ground can be cut to an accurate profile and will stand up long enough to enable a ring of segments to be laid directly against the ground and locked permanently in place by introducing a wedge shape segment to expand the ring against the ground. Because London clay is virtually impermeable no water enters despite the lack of grout outside the lining. With the Channel Tunnel, however, two factors meant that the London solution could not be applied directly. The first was doubt about the ability of the TBM cutter head to excavate the chalk to a profile of sufficient accuracy; the second was the need to provide a grout layer outside the lining to reduce the permeability around the unsealed linings.

The solution was to provide raised pads on the back of the linings to ensure that an annular gap was provided and that each segment was properly located prior to the lining being grouted. Expansion of the ring ensured that the ground was supported and the annular gap could be grouted up later in the construction process. As with the French linings, the UK linings were fabricated from high-strength reinforced concrete. Not all of the drives on the UK side were expected to be in good ground, however, and an alternative lining was necessary for coping with unfavourable conditions.

A small quantity of a second type of lining was procured as a replacement for the expanded concrete lining, which could be substituted without unduly affecting the overall rate of progress of the TBM. A bolted lining was indicated but a concrete lining would have needed a large annular grouting space and extra thickness to accommodate the bolts and would have led to a reduced internal diameter. A conventional segmental bolted cast-iron ring with grooves machined into its joints to accommodate neoprene waterproofing seals was therefore provided, thinner overall than the expanded concrete linings and thus not causing any reduction of the finished internal diameter. (Note that once the external diameter of the boring machine has been determined, the external diameter of the lining cannot change. Hence, any design requirement for a thicker lining must reduce the internal diameter.) The cast-iron lining was required to be grouted into the ground and to be built within a tailskin bolted to the rear of the TBM, but otherwise was segmented to enable it to mimic the concrete segments and thus allow the use of existing erection and handling equipment within the TBMs.

OTHER AREAS

Both the UK and French concrete linings were designed with articulated joints between segments to transmit the large hoop stresses developed within each ring from the ground loading. These articulated joints consist of a knuckle-type connection of either convex/convex surfaces (UK) or convex/concave surfaces (France), which resulted in the load transfer being localised to a line of contact. The effect of this localising of load on the segment is to induce very high compressive stresses at the point of contact with associated high tensile or 'bursting' stresses in the zone behind the joint. Layers of reinforcement steel are required in the joint to resist these tensile forces.

All the main TBM drives required provision to be made during construction of the tunnels for the subsequent connection of cross-passages and piston relief ducts. Not surprisingly in view of the different forms of lining used on either side of the Channel, this construction task was detailed in differing ways in the UK and French tunnels. In the UK service tunnel drives the standard concrete linings were replaced by a 6m length of expanded and bolted cast-iron linings, which incorporated framed openings on one or both sides of the tunnel into which removable panels had been placed to ease subsequent construction activities.

In the rail tunnels, where openings would only be located on the service tunnel side of the tunnel wall, a novel type of opening frame was designed with the twin purpose of minimising the amount of expensive cast iron required and also the impact on the overall rate of progress of the drives. These hybrid rings were developed to be formed partly of concrete segments and partly of cast-iron segments, of the same thickness and segmentation as the standard segments. Providing rings of this type enabled a cast-iron eye frame to be formed, segment by segment, bolted together to provide a frame within the concrete linings. At the centre of the frame were temporary removable panels to assist subsequent junction excavation.

The French method of providing openings from their service tunnel was to use composite bolted segmental linings of cast iron infilled with concrete, of the same thickness as the standard rings and which included framed openings complete with temporary panels. For openings in their rail tunnels, the French design was to make no specific provision during lining construction but subsequently to stitch-drill a circular opening through the linings prior to inserting a cast-iron frame into the hole so formed and packing up the void between lining and frame.

Apart from the main drives, all of the minor connecting tunnels were constructed by hand methods. Therefore linings needed to be both light enough to handle and build within confined spaces, yet strong enough to carry the substantial design loadings. On the UK side, the linings for cross-passages were designed as segmental bolted cast-iron rings with machined grooves in their joints to enable neoprene waterproofing gaskets to be used if required.

Where NATM was used to provide the primary temporary support structure in the permanent tunnels, a heavy final lining had to follow. The permanent support normally consisted of a plain concrete lining cast inside the primary lining incorporating the provision of a waterproofing membrane. The secondary lining was designed to carry the full permanent loading and assumed no assistance from the primary lining. Careful sizing and shaping of these secondary linings enabled them to be designed generally without the provision of steel reinforcement.

DESIGN MODEL

Following the studies of ground characteristics, and knowing the requirements of the construction process, the designers had obtained the two principal elements required to enable preliminary structural design to commence:

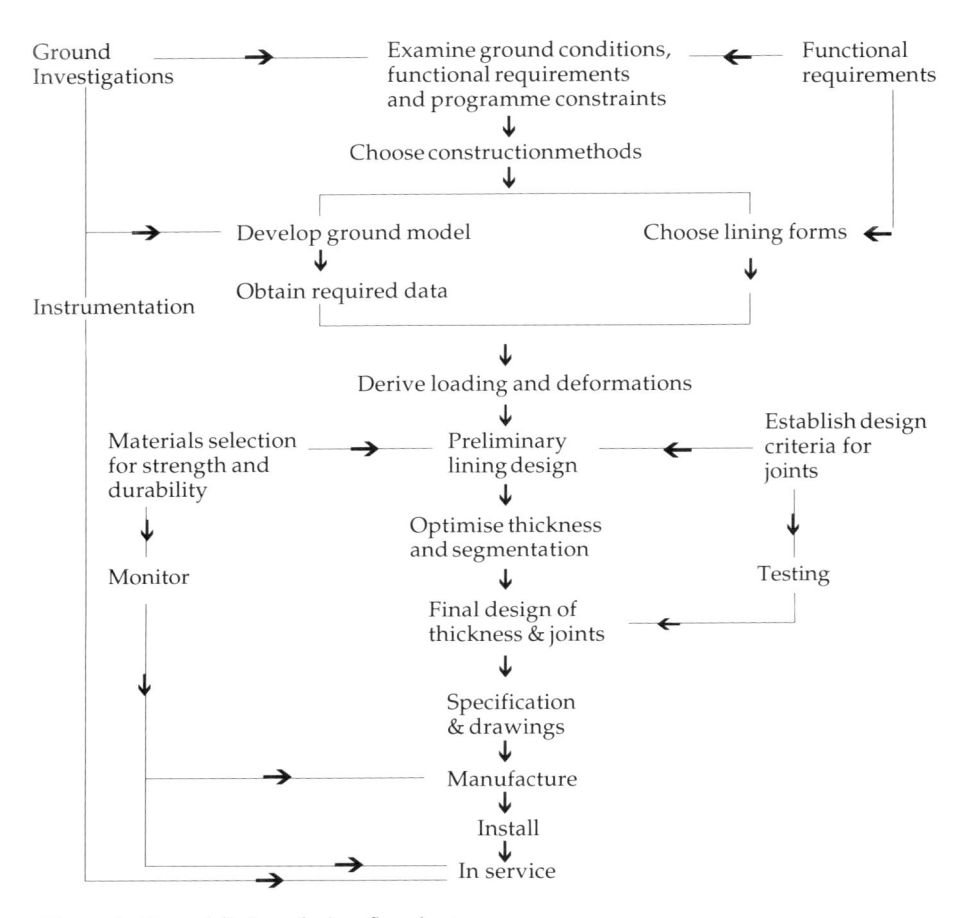

Figure 1. Tunnel lining: design flowchart

functional and construction requirements, and historical experience and site-specific ground investigation. A flowchart illustrating the overall design process for the tunnel linings is shown in Figure 1.

The first need was to derive an analytical method to determine accurately the loads that would be carried by the linings, and to predict the accompanying ground and lining deformations. Such a method required the development of a 'ground model' which could be used as a predictor and which could be calibrated against measured values derived from *in situ* instrumentation of completed test sections. The model also needed to take account of the alternative construction methods and the range of anticipated groundwater conditions.

Observations and measurements from previous projects had indicated that the behaviour of the ground during excavation and support had been to react with an immediate uniform movement (elastic) followed by a slow creeping movement (viscous). It was therefore possible, by assuming the ground to be a continuum, to derive viscoelastic relationships to predict the interactive behaviour between the ground and the lining.

The inclusion in the design equation of a term for ground deconfinement enabled account to be taken of the different TBMs and linings used by the UK and France on their main drives. In the UK, an open-faced TBM was used that did not allow face support to be applied to the ground between excavation and installation of the lining, and resulted in the ground movements being unconfined. Thus a deconfinement ratio of 100% was appropriate and indicated a reduction in the final loading applied to the lining. In France, however, the TBMs had a closed face which enabled a pressure to develop between the ground and machine, resulting in the ground being partly supported following excavation and before lining installation. These partial deconfinement ratios between 100 and 50% could be used to calculate the effect on final loads on the linings, which would normally be greater for French than for UK linings.

The effect of the speed of construction between ground excavation and the installation of the support had also to be evaluated.

By application of the model the effects of ground loading on the completed linings could be established. Lining loadings are also controlled by the action of groundwater and here two separate conditions need to be defined: where the completed lining is less permeable than the surrounding ground, and where the completed lining is more permeable than the surrounding ground. In the first condition the lining is assumed to be watertight, enabling the full hydrostatic pressure corresponding to the total head of water to build up on the outside. The French linings, by the use of waterproofing seals in their joints, fulfilled this criterion, and were designed for both the hydrostatic loading and the submerged ground loadings derived from the equation. This condition was termed the effective stress case. On the UK side, no seals were provided in the joints of the lining as the permeability of the ground was generally lower than for the French side. Control of groundwater inflow was effected by the cement grouting undertaken between the lining and the ground. Groundwater inflow was not however stopped, only controlled, with the result that the permeability of the lining was greater than that of the surrounding ground. Hence, the groundwater loading was applied to the lining through the ground and was taken as an externally applied load to the ground surface. This condition was termed the total stress case.

DESIGN PROCESS

Once the site investigation had been completed, tests could be undertaken on ground samples which allowed their physical and engineering properties to be defined. Thus, in conjunction with the analytical model previously developed, the ground loadings that would act on the linings could be established. Additional load factors were also derived to take account of the effects on one tunnel by construction of an adjacent tunnel. These factors were obtained following extensive analytical studies of all possible construction sequences. The factors used are given in Table 1.

Table 1. Load factors for tunnel interactions

Service tunnel	1.20
Rail tunnel	1.05
Intersections	1.4

All the linings were designed to comply with two ultimate-state loading conditions, as required by the construction contract:

- ultimate limit state with agreed partial load and materials factors using the derived ground loading
- ultimate limit state under full ground loading with partial load factors of 1.0.

The second condition enabled the Safety Authority to satisfy themselves that the permanent linings had been designed to a demonstrably acceptable ground loading.

A major element of the lining design was the detailing of the joints and in the concrete linings the need to obtain the economic weight and distribution of steel reinforcement. Following extensive design studies in both the UK and France a series of tests were undertaken on scale models of the joints, as test rigs were not available for full-size prototypes. Tests on the concrete joints took place in both UK and France, allowing each to reflect the slightly dissimilar detailing rules allowed by the national standards. The comparable results formed the basis for detailed design.

Tests on the cast-iron linings also took place in the UK and demonstrated that the proposed joint detail performed well, under load, and enabled enhanced design rules to be adopted.

One important concept adopted for the project was a requirement to design both UK and French works to common rules. This therefore entailed the development of a common set of partial load and materials factors based on national standards, together with a common definition of material design stresses. This area of design resulted in much debate between the UK and

French teams, but a set of project standards was eventually established and the principal criteria are given in the following tables.

Table 2. Partial load and materials factors

	Partial load factors		Partial materials factor γ_m				
	Ultimate	Service	Concrete Precast	In situ	Iron	Steel	Rebar
Ground	1.35	1.0	1.40	1.50	1.25	1.20	1.15
Water	1.35	1.0					

Table 3. Materials design stresses

Material	Design stress	
Concrete	$\dfrac{0.67 \times \text{characteristic strength}}{\gamma_m}$	
Iron (SNG)	Tension	0.2% proof stress
	Compression	1.1 × 0.2% proof stress
Rebar	Characteristic tensile strength/γ_m	

Table 4. Materials specification

Material	UK	France
Iron (SNG)	Grade 600/3	Grade 500/7
Precast concrete	C60	fc 45
		fc 55
In situ concrete	C40	fc 30

The design process for the TBM-driven tunnels defined the maximum thicknesses for the precast concrete linings in each drive, under the applied loading conditions (Table 5). In addition, it was sometimes possible to identify substantial lengths within each drive where lining thickness might be reduced from the maximum as a result of a decrease in the applied loading. The general economics of overall tunnelling logistics, however, militated against the manufacture and handling of too large a number of different thicknesses of lining and so the UK marine rail tunnels were the only drives to employ linings with different thicknesses.

Table 5. Lining thicknesses (mm)

	UK		France	
	ST	RT	ST	RT
Land	410	540	320	400
Marine	270	270/360	320	400

Durability of the tunnel linings was treated as being of prime importance, as their design life had been specified within the construction contract as 120 years. Conditions in the Tunnel were expected to be severe, from a corrosion risk point of view, and the designers needed to quantify these risks and offer mitigation measures, or provide additional protection. Particular risks were identified from the leakage into the Tunnel of waterborne salts, the high temperature/low humidity atmosphere, and the aerodynamic characteristics.

The first line of defence became the detailing of the works to eliminate the presence of free water surfaces inside the tunnels, ensuring that all percolating water was channelled into a closed drainage system for transfer to the sumps. In addition, the concrete linings were designed to ensure that cracking under load was eliminated and that the cover to embedded reinforcement was set at levels which would maximise protection against chloride ingress without detracting from structural performance. All materials used were tested extensively prior to manufacture to develop mixes that would provide the requirements of maximum density and minimum permeability and diffusivity while ensuring the elimination of any alkali silica reaction. The outcome was a concrete which set new standards in performance. (See also Chapter 5 for details of the manufacture and testing of tunnel linings at the factory on the Isle of Grain.)

A final and important aspect of the lining design was the verification of the design assumptions through the use of *in situ* instrumentation. This took the form of both structural and corrosion potential measuring stations at about 50 discrete locations distributed throughout the tunnel systems. These are being read at regular intervals and a long-term record of behaviour built up for comparison with anticipated performance.

5

Initial access developments

DAVID WALLIS

One could argue that the initial access developments were those that took place early in the last century. The original platform at Shakespeare Cliff, 16m above sea level, was formed in 1844 during construction of the railway between Dover and Folkestone and lies between the Shakespeare tunnels to the east and Abbots Cliff tunnels to the west. It was constructed of chalk rubble from the cliffs and its function was to provide a berm for the rail tracks.

In the late 1880s the platform formed the site of Shaft No. 2 (No. 1 being just west of the Abbots Cliff tunnel and No. 3 being just east of the Shakespeare tunnel) from which 1815m of 2.1m diameter tunnel were driven using Colonel Beaumont's full-face tunnelling machine driven by compressed air. This first real attempt at crossing the Channel is remarkable in that it was constructed in the chalk marl at the same level as the present scheme. In 1974 the old shaft, which had been backfilled, was located, re-excavated and lined, and the flooded tunnel was opened up.

The tunnel was in remarkable condition (see Chapter 2, Figure 1), much of it remaining just as it had been left over 90 years earlier with the grooves of the tunnelling machine still in the walls. One reason for its original flooding after abandonment was to prevent collapse. It had been unlined when excavated but local rock falls had been supported by lightweight, bolted cast-iron rings at about 3m centres with half tree trunks spanning between them. Where these had been installed the tunnel remained intact but in unsupported areas further falls had occurred.

About 800m of the Beaumont tunnel was opened up in 1974 to the point where the new tunnels were to cross, and galleries were excavated to install ground movement detection instruments ahead of the tunnelling machine. In the early 20th century, the site also provided the first attempt to win coal in Kent when three shafts were sunk to depths of 353m, 371m and 388m. The shafts were beset with difficulties due to water and coal was never mined commercially. The shafts were later partially backfilled and capped.

To understand the development of the works in 1986 to provide access to the tunnel construction it is necessary to go back in time to the last construction attempt in the 1970s. In November 1973, a Franco-British Treaty allowed the construction of 2km-long trial bores of the service tunnel, known as the Phase II Works, to begin in France and England. On the British side those works consisted of a road access tunnel from the top of Shakespeare Cliff to the platform at the cliff bottom beside the main Dover-Folkestone-London railway tracks, a steeply declined tunnel (drift) from this platform eastwards to the junction with the service tunnel some 50m below sea level, and a length of the service tunnel itself consisting of 2km of segmentally lined tunnel excavated by tunnel boring machine (TBM).

In January 1975 these works were cancelled due to the failure of the British government to secure the powers necessary for it to ratify the Treaty. At this time the TBM was installed and ready to start but had not commenced its drive. As a substantial amount of money had been spent installing ground instrumentation to enable the lining design to be optimised, approval was granted to drive 250m of the service tunnel through the instrumented zones.

The TBM with lining erected behind it could not be removed from the tunnel and it was left at the seaward end of the tunnel. In order to retain it in workable condition, it was necessary for the tunnels to remain accessible and to be ventilated and pumped. The road access tunnel, which had been supported only with arch ribs and timber lagging, was provided with a permanent lining of spray-applied concrete and the drift and service tunnel were lined with concrete placed *in situ* behind steel formwork.

ACCESS CONSIDERATIONS

While the scheme proposed by the Channel Tunnel Group/France Manche (CTG/FM) in 1985 was substantially the same as the 1970s scheme, and intended to make use of the previously constructed tunnels, significant differences affected the provisions for access and the way in which the project was planned.

First, the access tunnels which had been lined after cancellation were of even smaller dimensions than those necessary in 1974 and, secondly, the shortened construction programme demanded a greater rate of supply of materials and removal of spoil.

The CTG/FM proposal to government had envisaged that the 8km length of bored tunnels under land would be constructed from the portal at Holywell just north of Folkestone, and the undersea 22km length from Shakespeare. This proposal for the underland tunnels required a substantial working site at Holywell with cranage for offloading and stockpiling segments, a rail link to the main BR tracks at Dollands Moor, sidings, and a crossing of the M20 motorway. Spoil removal was planned by way of a conveyor around Castle

Hill to the terminal site where approximately 1 million cubic metres of chalk marl excavated from the Tunnel would be spread to level and raise the site.

During the passage of the Channel Tunnel Bill through the House of Lords in 1986, concern was expressed at the impact of the works at Holywell Coombe, which is classified as a Site of Special Scientific Interest. Furthermore the coombe forms a natural amphitheatre overlooking the town of Folkestone and it was considered that there may be objection to the impact of construction activities, especially as the programme was based on continuous seven-day, 24h working. A subsequent, more southerly, realignment of the tunnels at Holywell to avoid the sensitive coombe added to the difficulties of forming a working site by further restricting the area available.

ACCESS FROM SHAKESPEARE CLIFF

As a result of these issues the effect of driving the underland tunnels from Shakespeare was examined. Whilst there were clear advantages in removing the tunnel support logistics from Holywell, the doubling of the requirements to support all six drives from Shakespeare would obviously impose great demands on that site. Furthermore, the spoil which was required to fill the terminal area would arise at Shakespeare where there were already environmental objections to the disposal of about 3.75 million cubic metres of spoil from the marine tunnel drives.

When the construction programme was examined in detail it was recognised that the requirement for filling the area towards the eastern end of the terminal site could be delayed to allow the land service tunnel to be completed and for spoil to be returned through it from the marine tunnel drives. Moreover the terminal site in this area had a mantle of gault clay which would prevent any residual salinity from the marine chalk marl reaching the underlying aquifer.

At Shakespeare the initial plan had been to supplement the 1974 access adit with a second smaller adit to serve the three marine drives. The marine service tunnel drive could commence – as soon as parliamentary approval was given – from the chamber formed when the 1974 scheme was abandoned. The two larger rail tunnel drives required the construction of chambers for the erection of the TBMs and their backup equipment. The land tunnels required only the construction of chambers to receive those TBMs and allow them to be dismantled and removed. The facilities to support the TBM drives had been based on rope haulage in the access drift to supply tunnel lining segments and materials to the Tunnel and conveyors to remove spoil, similar to the approach adopted in 1974.

A re-evaluation of the programme based on both land and marine tunnels driven from Shakespeare showed that it would be necessary for five TBMs to be operating simultaneously. For the marine drives the service tunnel would be driven ahead to prove the ground for the main drives which would follow on completion of their preparatory works. Nonetheless, the three-year programme for the drives resulted in all three marine tunnels being under construction simultaneously for nearly two years.

For the land drives, emphasis would again be placed on the service tunnel to enable it to prove the ground. The two main drives would follow three months apart to allow completion of erection and commissioning of one TBM before commencement of the other. This arrangement resulted in virtual completion of the year long service drive before commencement of the last rail tunnel drive. Thus only two drives would be carried out concurrently.

To provide linings and materials to five concurrent drives it was found that a rack railway system would be more cost effective than rope haulage in the adits and it was decided that each drive should have its own dedicated track

Figure 1. Shakespeare Cliff site in operation, 1989. (Source: QA Photos)

to the surface. This were to be achieved in a single adit of elliptical cross-section, 12m wide by 7.8m high. At the bottom of the adit, three tracks curved seawards and two landwards in a bifurcated junction, with cross-connections to the three main tunnels.

For removal of spoil, the 1974 adit which entered the service tunnel at grade was extended downwards to pass beneath the main tunnels. A system of moving bed bunkers for each drive was installed beneath the rail deck in each marshalling tunnel with metering conveyors installed in short tunnels connecting with the main adit. In the latter a single conveyor with a capacity of 2400 tonnes per hour was installed to bring all spoil to the surface.

Considerable study was given to the main conveyor, as any problems with it would affect the entire tunnelling operation. It was concluded that provision of a second conveyor to guard against excessive downtime was unnecessary if duplicate motors were provided.

Figure 1 shows the Shakespeare Cliff site with work underway in 1989.

Reclamation

The decision to begin the land tunnel drives at Shakespeare Cliff instead of Folkestone meant that the surface logistics required for all six tunnels were considerably greater than those planned for the three marine tunnels only. The platform at the foot of Shakespeare Cliff would not have been large enough just for the marine drives, let alone for both the marine and land tunnels. It was planned to increase the area by reclamation using tunnel spoil placed behind a reinforced concrete sea wall.

During the passage of the Channel Tunnel Bill in 1986 there was considerable environmental objection to the proposal to enlarge the Shakespeare Platform. Concern related to the visual intrusion of the retaining wall, chalk mounding, the prevention of natural wave action against the cliffs allowing growth of vegetation which could result in 'the green cliffs of Dover', and the permanent covering of certain geological exposures and marine life on the foreshore. It was recognised that it would be necessary for some enlargement of the site to facilitate the works but there was an excess of spoil for that purpose and objectors did not want this to be deposited at Shakespeare. Over 70 other possible sites across Kent were examined but most had disadvantages in terms of the need for rail access, unloading facilities, structural requirements or their own environmental difficulties. It was finally accepted that placement at Shakespeare was the least damaging option.

The additional facilities required for the reversal of the direction of the land drives were able to make use of the extra reclamation to locate extended sidings, additional lining segment stockpile, locomotive and rolling stock maintenance and storage of tunnel construction materials.

Figure 2. Shakespeare Cliff site during tunnel construction showing lagoons. (Source: QA Photos)

A condition of approval to dispose of chalk at Shakespeare was that this should be carried out in enclosed lagoons to prevent dispersal of chalk fines over a larger area of sea bed (Figure 2). It was therefore decided to construct a peripheral sea wall consisting of two rows of sheet piles, 10m apart, infilled with concrete placed partly underwater to above tide level (+4.2m o.d.) topped with *in situ* concrete placed within removable steel formwork to above wave level (+7.0m o.d.). Above the wall the spoil was to be battered to the final landscaped level with the slopes protected against wave overtopping by concrete armouring. The wall was to be constructed westwards from near the portals of the access adits. Temporary crosswalls consisting of two rows of sheet piles with gravel infill were constructed at intervals to create lagoons so that a new enclosed lagoon was completed just prior to complete filling of the previous lagoon. For the initial six months of construction the marine service tunnel was built from the original 1881 site as the first small lagoon was filled. This made an additional working area available to support the startup of the first rail tunnel drives, and as the chalk cut by the TBMs came to the surface, so the working area increased to support the drives. The plan was to fill lagoons westwards on a broad front using a tracked elevated conveyor, the top 5m of spoil being compacted to permit erection of construction support buildings (for locomotive and rolling stock maintenance and stores) and storage areas for tunnel lining segments and concrete batching.

In the event, the slowness of the initial service tunnel drive resulted in less reclamation than expected when the main rail tunnel drives commenced and the site development had to be progressed differently in order to maintain tunnel production. Spoil was carried by dump trucks to more critical areas and reclamation carried out at a lower level to enable a greater area to be produced (Figure 3). Even so, some workshops and TBM laydown areas were forced to be located at the site on the cliff top.

ROAD ACCESS

It had always been planned that the main offices for management and administration of the UK tunnelling would be located at a site on the top of the cliffs at Shakespeare, where they had been in 1974. Access to this area, adjacent to the old Folkestone road, had been by way of the Aycliffe housing estate from Dover. However in view of the greatly increased volume and nature of construction traffic for the full-scale tunnel construction this route was obviously unsuitable. Alternative access was possible by way of a little used

Figure 3. Shakespeare Cliff site after completion of Tunnel, 1994, showing reclamation. (Source: QA Photos)

Ministry of Defence road from Farthingloe. This road had been used to serve the site in 1974 and, at that time, it had been improved with widened bends and laybys and a new junction with the A20 Dover-Folkestone road. By 1985, the road was considered inadequate to provide the capacity for the tunnelling works and an entirely new road was constructed.

In the planning for accommodation and transport of workers to the site it had been assumed that large numbers of the workforce would be brought to the site by contractors' buses from outlying areas and from the temporary village constructed at Farthingloe Farm adjacent to the A20. Even so, the number of car parking spaces at the clifftop was greatly underestimated and a large car park had to be constructed in an area of land beside the A20 with special buses serving the site at Upper Shakespeare.

A further complication for the upper site arose from the plan to improve the A20 trunk road between Folkestone and Dover. In response to representations from Dover District Council and the ferry companies, the Secretary of State agreed to the extension of the M20 from Folkestone to Dover in the form of a dual carriageway trunk road; the work would be carried out concurrent with tunnel construction and be complete by the opening of the Tunnel . The alignment of the new road at its eastern end was settled quite late in the process and it had not therefore been possible for allowance to be made in the Channel Tunnel contract for the fact that the road would run through the Upper Shakespeare working site. Discussions took place between Eurotunnel and its contractors on how to resolve the conflict in order to meet the obligation to complete the road in time for the opening of the Tunnel. This necessitated a sequential handing over of parts of the upper site and the construction of a Bailey bridge for access to tunnel completion works, for which the tunnel contractors were compensated. At the Folkestone end, provision for the new A20 had been included within the Channel Tunnel Bill itself where the road passed on a viaduct across the cut-and-cover tunnels at Holywell Coombe. This was because access to the Round Hill road tunnel necessitated access across the Channel Tunnel works.

ISLE OF GRAIN PRECAST WORKS

Because of the lack of space at the Lower Shakespeare Cliff site, the precast tunnel lining factory had to be remote from the pithead. After a prolonged search, the purpose-built facility was installed on the site of a former BP refinery on the Isle of Grain in the Thames estuary.

The advantages of the Isle of Grain were that the massive quantities of bulk materials required for segment production could be delivered by sea; and the finished precast units could be delivered to Shakespeare Cliff along existing freight lines.

Between October 1987 and May 1991, a total of 442,755 reinforced concrete segments ranging from 270 to 540mm in thickness were manufactured at the factory and delivered to the pithead. In all, the factory was responsible for making and keeping track of 268 different types of segment. Up to three train loads per day of finished segments were leaving the yard at peak times. At one stage, segment production was so far ahead of the requirements of the TBMs that there was insufficient room in the segment storage area at Shakespeare Cliff. A temporary storage area was created in Ashford.

Over 1 million tonnes of aggregate were used in segment production. Because of the demanding specification of the lining (see Chapter 4), both the fine and coarse aggregate chosen was processed granite mined in Glensanda, Scotland. A total 90,000 tonnes of pulverised fuel ash were included in the mix to improve the concrete's characteristics and to increase the impermeability of the finished product; 44,500 tonnes of steel were used for reinforcement.

The production lines worked on a carousel principle, with tailormade moulds moving through a series of work stations in a seven-hour cycle. First, they were fitted with a steel reinforcing cage. Then, the concrete, heated to a temperature of 20–25°C, was poured in, after which it was vibrated to consolidate the mix and remove air. After being smoothed off by hand, the moulds entered a curing tunnel to be steam heated. Leaving the tunnel, the segments were removed from the moulds, wrapped in thermal blankets to prevent cooling taking place too rapidly, and ferried to the storage yards. Here, they spent a minimum of 28 days before leaving for Shakespeare Cliff and the UK tunnel construction works.

Prior to the advent of the Isle of Grain, the accepted rate of failure for this type of precast concrete was 0.75%. Over the $4^1/_2$ years of the Isle of Grain's existence, the failure rate, in a typical week, was a remarkably low 0.25% or less.

Testing was extremely thorough. For example, the dimensions of segments were regularly checked in a special room at the end of the production line. The job was carried out by a robot. Run on software developed to check the turbine blades on Rolls Royce jet engines, it produced an analysis of 100 different measurements. Perhaps the most crucial of these was on the end of a segment where it butts up against another when a ring is formed. The allowable tolerance was just 0.1mm – the thickness of a piece of paper. Yet not one segment failed on this count.

There was also a highly sophisticated concrete laboratory, which carried out a comprehensive programme of tests on everything from the quality of the aggregate to the strength of the concrete. Ninety days after leaving the curing tunnel, the segments possessed a crushing strength of 90–100N/mm^2, probably the strongest production concrete in the world.

FRENCH ACCESS

On the French side, where ground conditions are not so favourable, less work had been completed in the earlier attempt before it was cancelled in 1975. An inclined access tunnel had been excavated and lined with cast-iron segments but progress had been delayed by severe water ingress. A diversion to try to regain time was under construction at the time of cancellation and, although the service tunnel TBM had been assembled on site, it had not been introduced into the tunnel. As a result, on cancellation, the protection works on the French side consisted simply of removal of any equipment and allowing the tunnels to flood, which occurred within about 24h.

The proposal to governments in 1985 had envisaged re-use of the 1970s works, with the main tunnels crossing the coast on a similar alignment. The intention was that the use of these previous works would reduce the time taken to gain access to the main tunnelling level. It was known from the 1975 experience that the heavily fissured and permeable chalk at that level would produce difficult tunnelling conditions. The main tunnelling concept at that

Figure 4. Sangatte shaft, 1994 (Source: QA Photos)

Figure 5. Fond Pignon disposal site, 1990 (Source: QA Photos)

time (1975) was based on open, full-faced TBMs with extensive pregrouting to reduce water ingress.

In order to minimise the effect of these conditions, an alternative scheme was developed that consisted of a large access shaft at Sangatte (Figure 4) surrounded by a grout curtain to prevent groundwater ingress. From here TBM erection and marshalling tunnel construction could be carried out in dry conditions. The main tunnel alignment was adjusted westwards to enable this to be accomplished next to the 1975 *descenderie* and a neighbouring burial ground. This also allowed the tunnel level to be reduced up dip. While this proposal was seen to cause slower start-up, the construction was much simplified and access to back up the TBM drives would not impede the advance.

Within the grout curtain the circular access shaft was 70m deep and 55m in diameter, thus embracing service and rail tunnels. The upper 21m of shaft were constructed using diaphragm walling techniques with the lower 47m underpinned by *in situ* reinforced concrete walls.

The French marine and land tunnels were driven from the Sangatte shaft (the land rail tunnels with one TBM driven from the shaft to the Beussingue Portal, turned and then driven back to Sangatte – see Chapter 6). The main tunnel working site adjacent to the shaft also included the precast yard for tunnel lining manufacture.

Whereas the geology on the UK side was expected to be uniformly dry at the tunnelling level, on the French side undulating bedding resulted in tunnels being situated in a highly fissured and very permeable chalk stratum.

While the original proposal was to grout the ground ahead of the open-faced TBMs, advances in closed-face machine technology (slurry and earth pressure balance types) resulted in their adoption, eliminating the need for ground treatment. The TBMs had to operate from the outset in 'closed' mode, in order to prevent water flooding the works, and the chalk spoil was delivered by the machine as a very wet slurry. As they advanced towards the UK, after about 5km they entered the chalk marl stratum, and could operate in 'open' mode, due to the practical absence of water. In both cases spoil was removed from the head of the TBMs by screw conveyor into rail-mounted skips. This slurried chalk spoil was discharged from the rail wagons into the bottom of the access shaft by a *culbuteur*, a device which rolled the wagons over, three at a time. The slurry was kept in its liquid state by constant agitation before being pumped overland to the disposal site at Fond Pignon (Figure 5). This was a huge reservoir, formed by the creation of a chalk rockfill dam 30m high across a dry valley on the low hillside adjacent to the shaft.

6

Tunnel boring machines

BOB MARSHALL

The philosophy of tunnel boring machine (TBM) design is to ensure a capacity to deal with conditions known to exist for the largest percentage of the tunnel drive, and to provide sufficient capability in the machine to overcome the unexpected, without impairing the efficiency in the known majority of conditions.

The chalk marl, which forms the vast majority of the ground along the Channel Tunnel route, is not considered to be difficult ground to excavate. In choosing the tunnelling system, TML had the benefit of a substantial amount of information on chalk marl as a tunnelling medium. Particularly relevant were the experience of Cross-Channel Contractors – the consortium of Balfour Beatty, Edmund Nuttall, Taylor Woodrow and Guy F. Atkinson – with the Priestley TBM on the UK service tunnel in 1974–75, the experimental work undertaken by the Transport and Road Research Laboratory in the Chinnor chalk-cutting trials with a McAlpine TBM in the mid-1980s, and the reviews of the geological studies/boreholes undertaken on the chalk marl along the tunnel route (see Chapter 2), published in 1985. All indicated that this stratum is generally stable, readily excavated with well-designed picks, as opposed to the disc cutters used in harder rock, and normally quite manageable in handling.

Thus, the dry and stable ground conditions anticipated by TML for 95% of the UK tunnel drives based on its assessment of the available geological information were not considered to present any special difficulties in the design and operation of the TBMs. It was assumed that the remaining 5% of the drive lengths would be lined with heavy-duty cast-iron rings and/or be subjected to ground treatment as a contingency against encountering seriously adverse ground conditions (e.g. fault zones, zones of high weathering). The TBMs had to incorporate in their designs the capacity to deal with such conditions.

Although ground conditions were not expected to cause problems, the UK tunnel drives did offer a unique challenge to both TML and the TBM manufacturers, through a combination of four factors:

- length of the tunnel to be excavated from a single access point, exceeding 20km in the case of the marine drives
- high speed of advance, up to 5m/h, required to meet the construction programme
- very high degree of reliability and consistency of operation
- ability to deal with, and contain, unexpected occurrences such as hitting an old site investigation borehole open to the sea.

SELECTION OF TBMs

Although the detailed design of the TBMs was to be the responsibility of the TBM manufacturers, TML was responsible for the conceptual design of the tunnelling system. The type of TBM, the tunnel lining type, and the form of the backup system were specified by TML based on the anticipated ground conditions and the aforementioned four factors.

The most suitable tunnelling system to cater for the dry, stable chalk marl anticipated for 95% of the drives was open, full-face TBMs, boltless expanded precast concrete linings (see Chapter 4) and a series of conveyors and muck skips for spoil handling. The use of an expanded segmental lining, which is erected against exposed ground at the rear of the TBM shield, has a great influence on the detailed design of the lining ring erection system, the grouting system, and other TBM functions. In addition, the length and thickness of the lining segments dictates the length and diameter of each cut, which influences the capacity and numbers of muck skips, the length of the segment and material supply/spoil disposal trains and, ultimately, the design of the TBM spoil conveyor system and length of the TBM backup gantries.

For the 5% bad ground contingency, it was envisaged that a spheroidal graphite cast-iron (SGI) lining would be used. Depending on the severity of the conditions, this lining could be built either against the exposed ground at the rear of the TBM or within a steel cylindrical tailskin bolted on to the rear of the TBM. Ground treatment could also be used to improve conditions in very wet ground. However, both these measures substantially reduce the achievable drive rates of the TBMs, and would have serious programme implications if used over a length of tunnel in excess of the 5% expected.

The extensive use of a boltless expanded segmental lining was advantageous from a programme viewpoint in that such linings are quicker to build than bolted linings in general, and are much quicker than bolted linings built within a tailskin. Although the annulus between the extrados of the rings and the cut ground required grouting, the use of bearing pads at the back of each segment to fully support the expanded rings in the predicted good ground conditions meant that grouting of the rings could be completely independent of ring building. Thus, the TBMs did not have to wait for the segments in the tunnel crown to be grouted prior to advancing, permitting faster drive rates to be achieved.

However, this speed advantage is sensitive to two additional activities in the production cycle, which may be forced upon the tunnelling gang by geological conditions:

- Firstly, if the rock is crisscrossed with minor fissures which cause it to separate into small pieces, then the small amounts of water in the fissures cause these pieces to fall out of the tunnel crown, and the voids which this creates have to be repacked to allow the tunnel lining to be expanded to its proper shape.
- Secondly, if the seepage of water becomes excessive, measures must be taken to control it. This means delaying the advance of the machine while cement grout is injected both to stabilise the lining rings as they are built, and to reduce the inflow of water.

Therefore, the combination of open (i.e. nonsealed) TBMs and expanded, boltless, ungasketed, precast concrete linings was ideal in terms of both the capital cost of TBM manufacture (closed machines are more expensive) and speed of drive in the anticipated dry, stable ground conditions.

The very high rates of progress required by the programme meant that the TBMs had to be able to build lining rings at the rear of the machine at the same time as the excavation was proceeding at the front. TBMs which only undertake one operation at a time generally use the last lining ring built to provide the reaction required during cutting to push the cutting head forward; only when cutting is finished can the forward thrusting rams (propel rams) be retracted and the next ring be built. To cut and build concurrently, the last ring cannot be used to provide this reaction, and therefore a 'gripper' section is required. This section uses large hydraulic rams to force bearing pads against the tunnel walls. The propel rams can then push off the gripper to force the main body containing the cutting head forward. At the end of the cut, the rams retract to pull the gripper section forward, thus exposing the ground required to build the next ring. The cut-and-build cycle can then be repeated. The use of a double 'telescopic' shield, i.e. a main body containing the cutting head and a rear gripper section, has a great influence on the detailed design of the TBMs, one result being that the TBMs are twice as long as single-action machines. On the marine rail tunnels, the rings were built 15m back from the excavated face. This distance can have a significant impact on stability in wet and blocky ground, as discovered on the UK marine drives.

How many machines?

As well as the selection of the type of TBM, other factors such as the number of TBMs, the timing of the drives to suit the programme requirements, and the drive directions all had to be determined.

The requirements of the construction programme could not be achieved by three single drives commencing at the land portal and terminating at the tunnel midpoints. On the UK side, three marine TBMs (two for the undersea portion of the rail tunnels and one for the same portion of the service tunnel) were therefore required, to commence driving at the 1974 workings at Shakespeare Cliff.

Because the marine service tunnel was to be used as a pilot tunnel for the marine running tunnels, it was essential for its TBM to be ordered as soon as possible after the contract commencement date. The greater the time lag between the marine service tunnel TBM commencing its drive and the commencement of the marine rail tunnel TBM drives, the greater use the information on ground conditions gained from the pilot drive could be in modifying the rail tunnel TBMs to suit the actual ground conditions.

Another control on the earliest possible ordering of the service tunnel TBM was a study into the risk of inundation which had concluded that this risk was very small, provided adequate probing was undertaken. It was decided therefore that throughout the length of the marine service tunnel drive, two 100m-long holes should be drilled at 4° to the longitudinal axis and at 3° elevation, at intervals of 75–80m. Downward and lateral probe holes were also required. Because of the need to stop regularly to undertake this probing, there was a possibility that the rail tunnel TBMs (which were not required to undertake investigative probing) would progress at a faster rate than the marine service tunnel TBM. Thus, the marine service tunnel TBM needed to start its drive well in advance of the marine rail tunnels to avoid the rail tunnel drives catching up. In the event, there was nearly 18 months' lag between the commencement of the marine service tunnel and rail tunnel drives.

The land tunnels were all driven uphill from Shakespeare Cliff. At least two land TBMs would have been required, one for the 5.76m-diameter service tunnel and one for the 8.84m-diameter rail tunnels (both diameters external/ unlined). Despite the drives being only 8km long, programme considerations relating to access for fixed equipment installation dictated that two rail tunnel TBMs were required, making a total requirement of *six* TBMs on the UK side. The short Castle Hill tunnels were driven by road header, not TBM.

On the French side, the shorter land drives of 3km permitted the use of only one land rail tunnel TBM, this single TBM undertaking both land drives by reversing its drive direction on completion of the first drive. The French side thus required *five* TBMs in total: three main rail tunnel machines (two marine, one land) and two (marine and land) for the service tunnel.

SPECIFICATION OF UK TBMs

Prior to issuing the enquiry documents to tenderers, the type of TBMs, their number, the delivery programme, and the factors influencing their detailed design (lining type, width of segments, method of spoil removal, capacity of muck skips) had all been determined. Thus the TBM manufacturers, although responsible for the detailed design of the TBMs, had little input into the conceptual design. Although the supply was determined by open tender, to avoid putting all one's eggs in one basket, it was decided that the order for all four rail tunnel TBMs would not be placed with only one supplier.

A general performance specification had been issued with the first tender enquiry for all six machines in February 1986. A similar specification was issued with the invitation to tender for the rail tunnel TBMs in April 1987 but it differed from the earlier one in that it was written specifically for the rail tunnel machines, and included requirements for inundation protection, fire protection and enhancements for pumping and ventilation.

General requirements

Manufacturers were invited to bid for the supply of the following machines, complete with all backup equipment (all diameters external):

- two marine rail tunnel machines at 8.5m diameter for 25km of drive
- two land rail tunnel machines of 8.84m diameter for 10km of drive
- one marine service tunnel machine of 5.38m diameter for 25km of drive
- one land service tunnel machine of 5.76m diameter for 10km of drive.

The machines were to be fully erected and tested in the works prior to delivery to site. The works tests were required to prove all TBM functions, including the erection of a lining ring within the specified time. All machines were to be designed and constructed so as to minimise routine maintenance and adjustment and to operate effectively under onerous conditions without major overhaul.

Performance requirements

The manufacturer was required to demonstrate during commissioning that the rail tunnel TBMs had the ability to sustain the full cycle of excavation, machine advance and segment erection at the rate of 16 cycles in 6h for three consecutive cycles, i.e. a rate 4m/h. The service tunnel TBMs had to be capable of 20 cycles in 6h, i.e. 5m/h.

Safety

The machines were to be equipped with a fire protection system consisting of local foam sprinklers and a general high-expansion foam system installed at the rear of the TBM backup gantries. The machines had to be capable of preventing inundation from fissures and dealing with the possibility of encountering a borehole which had been inadequately backfilled to the extent that it was effectively open to the seabed. It was not required that the machines work through and past a borehole without interruption. However, they were to contain the situation and prevent inundation or major flooding of the tunnel while the borehole was dealt with. Inundation prevention was to be achieved by the ability to quickly seal off the tunnel face in an emergency, and a means of sealing the annulus between the body of the machine and the cut rock and sealing between the gripper housing and the cutting head body of the machine. For the marine drives, support for the ground was to be provided between the gripper housing and the last ring built. These requirements were actually achieved on the marine rail tunnel TBMs by the following methods:

At the face: On hitting the emergency inundation button, the front section of the primary conveyor and hopper were withdrawn from the TBM head, and the conveyor access hole was closed. This whole process took only 15s. At the same time, knife seals expanded to block the annular gap around the TBM body. Both sets of gripper units were put into operation and gripper pressures were increased to withstand the 6600t force exerted by the maximum hydrostatic head acting on a sealed TBM.

At the rear: Between the rear of the TBM tailskin and the last completed expanded segmental ring built, up to 3m of ground could be exposed for the purposes of ring building. To prevent inundation via this exposed section of ground, shutters located within the TBM tailskin were planned that could be extended out of the tailskin on the auxiliary thrust rams and be positioned to seal off the exposed ground. However, due to the problems with overbreak and the use of a flexible steel hood over the ring-build area, the idea of using inundation shutters was abandoned.

None of the inundation measures provided on any of the TBMs was ever called upon, but the knowledge that they were there was certainly comforting.

Duty

The machines were required to operate in lower chalk, gault clay and mixed faces, although most of the drives were expected to be in the chalk marl layer of the lower chalk. Some sections of the drives were expected to be in zones that had been subjected to ground treatment. The tenderers were asked to refer to the extensive published information on the geological and ground conditions along the proposed route, and were expected to assess this information themselves for the design of the machines. The machines and their components were to be capable of a minimum penetration rate of 6m/h in all materials to be encountered. All major parts were to have a design minimum life of 20,000 operating hours. Access for assembly of the machines on site was to be through a 9m-diameter access shaft at Shakespeare Cliff upper site.

Electrical system

The machines were to conform to Section 7.25 of BS 6164 and CP1015 and to the IEE Regulations, 15th edition. The control equipment enclosures were to conform to BS 5490 IP65. This standard requires enclosures to be dustproof and to be able to withstand a water jet sprayed at any angle for a specified period. In view of the subsequent problems with the electrics on all marine TBMs due to saline water running onto electrical enclosures, it is arguable that IP67, catering for equipment that is fully submerged, would have been a more appropriate standard to use. The voltages used on the machines were to be power at 415V, control and lighting at 110V, and supply at 11kV. The machines were required to have a flexible trailing cable contained on a reeling drum capable of extending a minimum of 400m behind the rear of the backup gantries.

Hydraulic system

Hydraulic power was to be supplied by fully duplicated systems which were to be capable of operating independently, thus providing increased reliability. Hydraulic oil coolers were required on the machines to maintain acceptable operating temperatures, with the cooling water supplied through a hose reel from the 100mm-diameter supply pipeline.

Rotation and propulsion

Motive power was to be electricity, with all functions arranged to be failsafe. Multiple motors were required to drive the cutting head, which was to be dual speed and capable of cutting in either direction with equal power. Dual direction rotation was specified to produce even pick wear and as a means of correcting 'roll' of the TBM body. The method of propulsion was to be such that it could be advanced whether the lining was erected immediately behind the face or not. It was recommended that forward thrust be derived from a reactor ring or pressure pads bearing on the surrounding ground. This gripper pressure had not to exceed 21bar in normal conditions to avoid appreciable ground deformation. A 50% increase in this pressure was permitted in an emergency situation, e.g. potential inundation. Suitable provision was to be made for propulsion under all ground conditions.

Cutting head and shield

The cutters were to be readily removable and/or interchangeable with other types of cutter. Picks were the preferred type, although provisions were to be made for installing disc cutters should harder rock be encountered within the drives. Gauge cutters had to be adjustable to vary the cutting diameter of the machine by up to an additional 100mm. The ability to retract the cutting head from the face for inspection and changing/adjustment of cutters was required. Openings in the cutting head were to be provided for drilling boreholes into the face for forward exploration and pressure grouting of the ground ahead. Holes were also to be provided to permit lateral probing and ground treatment. The machine was to incorporate a fully cylindrical body to extend from the cutting head up to the segment erection position. Provision was to be made for the addition of a tailskin in which bolted linings could be erected should ground conditions dictate.

Exploratory probes

Two hydraulic drilling machines were required in each TBM head. They were to be capable of drilling a 50mm-diameter hole, through the probing/ground treatment openings described, of 100m in length within a 6h period.

Guidance

A ZED guidance system type TG 260 was to be installed in each machine to maintain line and level. This system was used in conjunction with a laser, which was set up just behind the last TBM gantry and close to the tunnel wall (to avoid the beam being blocked by the TBM gantries and equipment). The laser position was regularly moved forward down the tunnel as the TBM advanced. The laser beam was set up to a predetermined location and alignment so that it could be received on a screen mounted at the rear of the TBM cutting head. The ZED system's computer calculated the actual location of the laser beam on the screen and compared it with the designed tunnel alignment stored in the ZED system. A display of discrepancies between designed and actual cutting head position allowed the TBM driver to make continuous alignment adjustments by steering to zero. On the marine rail tunnel TBMs, the TBM driver used a joystick to steer the TBM head via the arrangement of propel rams. The use of the propel rams for steering meant that the driver could effect continuous corrections to the alignment throughout the cut.

Pumping systems

For the marine rail tunnel TBMs, a duplicated system of automatically activated inrush pumps was required, to be capable of pumping 300l/s through two 350mm-diameter pipelines 1500m long to the tunnel pumping sump. An emergency battery powered pumping facility, capable of pumping 150l/s for a minimum of 2h, was also required.

Spoil removal and material supply

Provision was to be made for the removal of excavated material from the face to the rear of the machine into 14m^3 capacity muck skips. Trains of 140m in length, with equipment for unloading and storage of two rings of segments and associated materials, had to be accommodated within the TBM backup gantries. No further guidance was provided in the specification as to the arrangement of trains within the TBM gantries, i.e. twin or single tracks within the gantries, loading/unloading arrangements.

Lining erection

The tenderers were advised that alternative linings for the tunnel had been designed in precast reinforced concrete and spheroidal-graphite cast iron. The machines had to be capable of handling all lining types and drawings of the lining types were appended to the specification. Regarding the precast concrete linings, which were intended to be used for an estimated 95% of the drives, Table 1 summarises the information passed to the tenderers (and see Chapter 4). The expanded concrete linings had, by necessity, to be built and expanded against exposed ground at the rear of either the gripper section or the tailskin, if one was fitted. Although it was intended that full cast-iron rings would also be built against exposed ground, it was also possible to build cast iron within the tailskin if ground conditions so dictated. The tenderers were advised of these requirements, and were informed that the segment erectors had to be capable of performing and operating simultaneously with the tunnel boring and spoil removal operations. These erectors were required to be capable of performing one complete ring-build cycle in 18 min in both service and rail tunnels, i.e. an advance rate of 5m/h.

Operation and maintenance

The essential, very high degree of reliability was to be achieved by, *inter alia*, duplication of systems and redundancy within systems. Standard components were to be used for maximum interchangeability of parts between machines. Easy access was to be provided to parts thought likely to require replacement during the drive.

Table 1. Precast concrete lining requirements

	Segments per ring	Segment thickness (mm)	Ring ext. dia. (m)	Ring length (m)	Weight per ring (t)
Service tunnel					
Marine	6 + 1 key	270	5.34	1.5	15.98
Land	6 + 1 key	410	5.62	1.5	25.41
Rail tunnels					
Marine	8 + 1 key	270/360	8.32	1.5	25.37/33.15
Land	8 + 1 key	540	8.68	1.5	51.08

The ventilation system in the rail tunnels was to deliver a minimum of $300m^3$/min of fresh air to the rear of the TBM through a 2m-diameter ducting. The cassette system for extending the ducting and the ventilation system within the TBMs were to be supplied by the TBM manufacturer.

The tenderers were required to provide labour, plant, materials and supervision for offloading and transporting the TBM components to the TBM erection chambers, site erection of the machines, and commissioning of the machines. In addition, they were to allow for operator and maintenance personnel training at the works and on site.

DESCRIPTION OF UK TBMs

Service tunnel TBMs were ordered from James Howden and Co. in November 1986. The marine rail tunnel TBMs were ordered from Robbins Markham Joint Venture in August 1987, and the land rail tunnel TBMs from Howden in December 1987.

Service tunnel machines

Both land and marine machines were designed to the same specification, so that the major components of each machine were the same, permitting the interchange of spare parts. The cutting head comprised four radial arms, each fitted with two rows of rocking picks, totalling 68 picks on the marine service tunnel and 76 picks on the land service tunnel. The cut spoil was collected by loading blades and dropped through a central aperture into a hopper which discharged onto a spoil conveyor. The front section of the conveyor was capable of being withdrawn from the head in the event of inundation.

The overall length of the TBM was 13m, consisting of the main body and cutting head assembly which telescoped with the rear gripper shield. The TBM was steered by a combination of the propel rams and by pairs of front and rear hydraulic steer shoes. Prior to the cut commencing, these steer shoes were extended to the walls of the tunnel and adjusted to provide the correct angle for the cutting head to be shoved forward by the propel rams. Cutting head torque and roll of the TBM were controlled by diagonally opposed hydraulic cylinders fitted between the rear of the main body and the front of the gripper section. A 40m-long beam was towed behind the TBM to carry the segment handling and erector systems.

Ring building was undertaken immediately behind the gripper section. The three lower segments were placed using a small rack-mounted crane. The three upper segments and key segment were lifted into position by a hydraulically operated 180° segment erector. Segments were attached to the crane/erector by a quick-release pin located in a cast-iron pot, cast into each segment. Prior to the ring being expanded by driving home the wedge-shaped key segment, the upper segments were held in position by hydraulically operated, retractable 'build bars'. To the rear of the ring-build area, backup gantries spanned the single 900mm-gauge offcentre railway track access to the segment unloading area. These gantries provided housing for the single spoil conveyor, the hydraulic power packs, grouting equipment, and ventilation equipment.

The total length of the TBM and backup gantries was 180m. This length permitted a 102m-long train to be accommodated on the single track within the gantries. Twin 900mm-gauge tracks provided access from the marshalling area at Shakespeare Cliff to the rear of the TBM gantries. Each train was long enough to carry two complete lining rings on flatbed wagons at the front of the train, together with the ten muck skips required to remove the spoil generated by two 1.5m advances of the TBM (approximately 130m^3 using a 1.9 bulking factor). The spoil conveyor was of a telescopic design to permit even loading of muck skips, allowing the trains to remain stationary during loading.

Cement for grouting was originally transported in bulk for use with a bulk cement silo and Schwing grout pump, both located on the TBM gantries. Although this grouting system performed reasonably well in the dry conditions found in the land service tunnel, the system was replaced by pan mixers, mono pumps and bagged cement in the marine service tunnel.

Figure 1. Marine rail tunnel north TBM in UK crossover cavern. (Source: QA Photos)

Marine rail tunnel machines

The two identical (but handed) Robbins/Markham TBMs comprised a cutting head with a total of 276 picks mounted on eight radial arms, linked telescopically to a rear gripper section by a series of angled propel rams (Figure 1). Twelve two-speed reversible electric motors were used to drive the cutting head at speeds of either 1.67 or 3.33 rev/min. The angled arrangement of the propel rams was designed to provide accurate roll and steer control to the TBM. The gripper section housed two pairs of gripper shoes (bearing pads); generally only one pair was used in normal operation. The TBM was also fitted with stabiliser shoes in the forward shield (main body) and auxiliary thrust rams at the rear of the gripper shield. These two sets of rams could be combined to assist the propel rams to pull forward the gripper shield after completing a cut.

Spoil handling was carried out by three belt conveyors. As in the marine service tunnel, the primary conveyor was fitted with a retractable front section to be withdrawn in the event of inundation. The third conveyor, or the shuttle conveyor as it was called, comprised a fully reversible belt conveyor mounted on wheels which ran on a short track above the parked muck skips. By travelling up and down this track and reversing the belt direction, spoil could be discharged evenly into either of two trains on the double 900mm-gauge railway tracks passing under the backup gantries. As with the service tunnel, each train carried sufficient segments and muck skips (11–12 skips per train) for only one 1.5m advance. However, unlike the service tunnel, two trains could be accommodated side by side within the TBM gantries because of the larger tunnel diameter.

Lining segments were carried into the tunnel on flatbed wagons ahead of the muck skips. The three lower segments (one invert and two haunch segments) and the key were transported forward to the two lower segment erectors using a low headroom gantry crane and a low-level segment conveyor. The remaining segments were lifted to an upper segment conveyor using an upper gantry crane and were then carried along the crown of the tunnel to the upper segment erectors. The upper and lower segment conveyors could accommodate two full rings. The upper erectors consisted of two carriages operating on a common ring beam, each capable of being operated independently to position the segments correctly. Hydraulic build bars maintained the position of the segments until the key segment was inserted and driven home using the key ram located in the gripper shield.

A special feature of the ring-build system was that, by the use of four separate segment erectors and by setting the invert segment one ring ahead, four of the remaining seven segments were designed to be placed within 3 min of the start of the 15 min ring-build cycle. Subsequent parallel placement of the upper three segments followed by keying up was intended to maximise float time, giving the operators latitude to improve significantly build time with experience. In the event, this system of ring building never achieved its design potential.

The original grouting system was similar to that in the marine service tunnel, but with the addition of a ring main through which pumped grout, kept alive for up to several hours by the use of retarder (Conbex 802), could be tapped off at various locations for grouting behind invert, sidewall and crown segments. An accelerator (Conbex 803) was automatically added in computer-controlled quantities at the grouting nozzle (at the point of injection through the grouting hole in the segments) so that the grout would remain fluid for just enough time to fill fully the annular void behind the lining, i.e. the setting time of the grout could be accurately controlled depending on the circumstances. The accelerated setting of the grout prevented washout by groundwater and provided immediate support to the ground and stabilisation of lining rings.

As with the marine service tunnel, the use of bulk cement and a Schwing pump was discontinued in favour of colloidal mixers, mono pumps and Pozament GP3 (a mix of 1 part cement to 3 parts pulverised fly ash), which was supplied bagged owing to problems with dampness affecting the bulk GP3, together with other considerations. The Fosroc designed grouting system of ring mains, albeit in modified form, with computer- and flowmeter-controlled retarder/accelerator addition, was used throughout the drives. The effectiveness of this grouting system in reducing water ingress into the tunnels, and the consequent obviation of the need for fluffing up at joints (i.e. temporary sealing at joints between segments to prevent leakage of grout) speeded up grouting and reduced labour costs on grouting operations. The overall length of the TBM and gantries was 250m, with an overall weight of 1500t.

Land rail tunnel machines

The two Howden TBMs also had a cutting head arrangement of eight radial arms, fitted with a total of 268 picks. Like the service tunnel TBMs, the propel rams were parallel to the drive direction, steering being effected by pairs of front and rear steer shoes. A single continuous-belt conveyor carried the spoil a distance of 200m to the discharge area. Loading of the seven muck skips per train, which were sufficient for half-a-ring advance, was undertaken by a travelling side-discharge conveyor (tripper) which ran within the main conveyor. Two trains could be accommodated side by side within the gantries on twin 900mm-gauge railway tracks. However, unlike the Robbins/Markham machines, the twin tracks within the gantries were not fitted to the concrete invert segments but were fixed onto the continuous closed deck of the backup gantries. All were towed by the advancing TBM. Trains gained access to this deck via a 25m-long ramped section of track towed at the rear of the machine.

Two independently powered, fully rotational 360° segment erectors were mounted one behind the other on a 14m-long beam spanning from the rear of the gripper section to the first backup gantry. Unlike the Robbins/Markham TBMs, all segments were delivered to the build area via a single path. Segments were unloaded from flatbed wagons by two electric travelling cranes which carried the segments forward and placed them on a segment magazine in the tunnel invert. At the build area, the segments were lifted off the magazine into their final position by either of the two erectors. The segments were installed in a two-ring build configuration, the three lower segments of the leading ring being built after the five upper segments and key of the previous ring had been positioned and expanded. As with the marine rail tunnel TBMs, the ring-build area was located at the rear of the gripper section, approximately 15m back from the tunnel face.

The grouting system of ring mains and the use of retarder/accelerator was similar to that initially employed on the marine rail tunnels. However, unlike the marine rail tunnels, the grouting locations, the grout pumps and the ring main arrangement did not require modifying to cope with wet conditions (conditions in the land drives were generally dry).

The overall length of the TBM and backup gantries was 280m, with a weight of 1700t.

PERFORMANCE OF UK TBMs

Performance in marine service tunnel

The marine service tunnel was the first of the UK drives to commence. The TBM began its journey in November 1987 from the position where the 1974 TBM drive had been abandoned. The TBM subcontract required the manufacturer to demonstrate that the machine could achieve a cycle time of 18 min (20 cycles in 6h). In the returned tender documents, the manufacturer guaranteed this performance by the concurrent building of the bottom half of the leading ring and top half of the previous ring, but with a proviso that the bore was regular, circular and without overbreak. This guarantee was also subject to water ingress not exceeding 10 l/min.

The early period of the drive was in dry conditions similar to those encountered on the 1974 drive. Several problems were encountered, however, the most notable being the failure of segment lifting pins on the upper erectors. Because of these teething problems, the theoretical cycle time was never achieved in the dry conditions and the concurrent building of two rings was never attempted. The rings were built one at a time with the lower erectors (cranes) placing the bottom three segments, and the upper system (segment loader and erector) placing the remaining three segments and the key segment. Maximum progress was 75 rings in one week.

Difficult ground conditions started to influence progress in late March 1988, with water ingress causing breakdowns of the electrical system on the TBM and overbreak affecting the ring-build quality and cycle times. This overbreak and water ingress meant that the standard cavity grouting procedures (grouting the invert at the front and grouting the upper half of the ring 25–30 rings behind the face) could not always be followed, and at times it was necessary to grout the rings fully within a few rings of the face. This forward grouting was not envisaged in the design of the TBM and the limited access in the build area owing to the presence of the conveyor bridge and the erector meant that ring building and grouting the upper half of the ring could not be carried out simultaneously. Grouting of the rings at this location further hindered progress.

Despite the TBM manufacturers' performance guarantees regarding the ability to deal with 10 l/min water ingress, it was clear that the TBM electrical systems could not cope with saline water ingress of even less than this amount, judging from the frequency of electrical breakdowns. It was therefore agreed to improve the waterproofing to electrical enclosures, pendant controls, electrical pickups, etc. during maintenance shutdowns.

The TBM as delivered to site was not capable of building a bolted cast-iron lining within the tailskin, because of the ram arrangement and the inability to grout the rings immediately behind the tailskin without grout penetrating around the tailskin and gripper section. The continued problems with ring building against the exposed and overbroken ground at the rear of the tailskin resulted in the decision being made to modify the TBM to build cast-iron rings within the tailskin. The modifications included the fitting of spacers to the rams to enable both the building of a bolted cast-iron lining within the tailskin or a concrete lining outside. The annulus between the cast-iron ring and the ground in the invert was filled with pea gravel, grout being injected 1–2 rings later to prevent grouting in the tailskin.

This solution was only used for a total of 65 rings before a flexible hood system was devised to speed up progress. This hood permitted the building of expanded concrete rings outside the tailskin by providing support to loose ground in the crown of the tunnel. By reducing the amount of overbreak, the need for the difficult, time-consuming and hazardous operation of packing voids behind segments prior to expanding rings was similarly reduced. A further benefit of the introduction of the hood was that fewer instances of ring

Figure 2. Marine service tunnel: formal breakthrough event, 1st December 1990. (Source: QA Photos)

distortion occurred owing to movement of packing when rings were expanded. The hood consisted of steel trailing fingers which fitted between the bearing pads on the rear of the segments so that the fingers would not be trapped when the rings were expanded. The fingers were fixed to the rear of the tailskin and spanned across the build area to the last completed ring. Stainless steel infill plates spanned between the fingers to prevent smaller blocks from falling into the ring-build area below.

The average progress in the wet and blocky ground increased to about 115m per week following the waterproofing of the electrics and the fitting of the flexible hood. However, due to the continuing need to pack overbreak the average cycle time was limited to 68 min. These difficult ground conditions persisted until December 1988, by which time only 3.5km of the drive had been undertaken. Once out of the wet and blocky ground, cycle times were reduced to about 30 min with a subsequent increase in average progress to 200m per week, and a peak performance of 293m in one week.

First contact was made with the French service tunnel on 30th October 1990, through a 50mm-diameter probe hole. The last permanent ring was installed on the same day, after which the TBM was driven offline to a position adjacent to the French TBM. Formal breakthrough of the tunnel (Figure 2) was achieved on 1st December 1990 through a short man-sized heading, virtually three years to the day after the drive commenced.

Following the removal of the TBM backup gantries, full-sized breakthrough was achieved using hand excavation and SGI lining rings. The UK TBM was progressively backfilled with concrete in its final resting place as this operation proceeded. By contrast, the French TBM was cut up and removed so that only its outer skin was left in position by the time full-sized breakthrough was achieved.

Performance in land service tunnel

The 8.1km land service tunnel drive commenced late September 1988. Generally, the drive progressed well in the anticipated dry conditions, achieving a post-commissioning average drive rate of 161m per week. Because of the generally dry conditions, the grouting system of bulk cement/Schwing grout pump did not require replacement as had happened on the marine tunnel drives, where dampness created problems with the use of bulk cement. The invert and haunch segments were grouted at the build area using grout from a paddle mixer, and the remainder of the segments were grouted 8–10 rings behind the build area using the Schwing pump.

Only over the last 600m of the drive, where glauconitic marl entered the face, did the ground conditions prove to be difficult. Blocky and wet conditions caused problems with ring building (as described for the marine service tunnel) and with cutting. The mixture of the chalk marl/glauconitic marl and water produced a very sticky material which clogged the cutting head hoppers. If this material was allowed to harden it became extremely difficult to remove, and on occasions the miners were required to enter the head to clear blockages using hand tools.

The final low-cover section to breakthrough at Sugar Loaf Hill was lined with SGI rings. The first breakthrough of a UK TBM drive occurred on 9th November 1989, 405 days after commencing the drive. The land service tunnel TBM was later refurbished to be used on the Brighton sewers sea outfall tunnel.

Performance in marine rail tunnels

The successful progress of the marine rail tunnel drives was considered critical to the success of the project as a whole. Considerable attention was therefore focused on the performance of the TBMs by the manufacturers, TML, Eurotunnel and the Maître d'Oeuvre. It was under this intense pressure to perform that the marine rail tunnel north and south TBMs commenced their drives in mid-March and mid-June 1989, respectively.

As a result of the experience gained in the wet and blocky ground in the marine service tunnel, provisions were made to fit a flexible hood (i.e. the steel fingers and stainless steel infill plates) over the ring-build area on the marine rail tunnel TBMs. In addition, a protective steel canopy was fitted over the top of the TBM gantries, close to the tunnel crown, to protect the TBM driver's cabin, the hydraulic power packs, and the TBM electrics between the ring-build area and segment/materials unloading area from the ingress of saline water.

The north TBM drive commenced in stable, dry ground and more or less achieved the programmed learning curve rates. The programme allowed for 18 weeks of steadily increasing production prior to reaching the target drive rate of 170m per week for the expected (from the pilot drive) wet and blocky conditions which existed between the 1 and 5km points of both drives. After the 5km point, the programmed rate increased to 235m per week for the remainder of both north and south drives. The peak production during the initial 1km of the north drive was 114m. The only problem encountered in the dry ground was the inability of the cutting head to rotate in both a clockwise and an anticlockwise direction. This was due to problems with the spring-back mechanisms on the picks, which resulted in the tips of the clockwise picks being dragged off when the head rotation was switched to an anticlockwise direction.

Prior to entering the wet and blocky ground, engineers from both TML and Eurotunnel were reasonably confident that the TBMs, as modified by the pilot tunnel experience, would be capable of coping with the expected conditions. Unfortunately, this confidence was misplaced: as soon as the north TBM encountered the wet and unstable ground in early July 1989, it suffered from major ring building problems and repeated breakdowns of electrical equipment.

Because of the larger diameter of the rail tunnels, overbreak in the crown of the build area was of a far more serious nature than that experienced in the service tunnel. This was due to the much larger zone of influence, leading to much larger blocks being affected by the excavation. The problem was exacerbated by the crown of the rail tunnels being higher in the geological series than the service tunnel crown, thus being closer to the zone of weathering extending down from the seabed. This meant that the permeability of the ground in the rail tunnel crown was much higher than in the service tunnel crown, so water inflows into the ring-build area were much greater, exceeding 100l/min/m length of tunnel at times.

The whole situation was made worse by the frequent electrical breakdowns caused by the saline water ingress, which resulted in equally frequent stoppages of tunnel driving. These stoppages allowed the unsupported ground between the excavation face and the ring-build area to relax, again made worse by the use of a gripper section which increased this unsupported length of ground. Thus, the slower the progress, the more the ground relaxed, and the worse the ground conditions became, leading to even slower progress: a downwards spiral of worsening ground conditions and progress.

For eight or so weeks into the wet and blocky ground, TML persisted with the use of an expanded concrete lining. Although the fingers spanned over the build area, the weight of loose blocks on the fingers was so great that they deformed to such an extent that the rings could not be expanded into their final position. In this situation, the only solution was to use pneumatic hand tools to break up the blocks behind the fingers to release the weight on the fingers. This process created voids behind the fingers, up to $3m^3$ on occasions, which then had to be hand packed with timber and cement filled bags to provide passive resistance to the expansion of rings.

Although progress averaged 52m per week during this period, through the use of the expanded concrete lining, ring-build quality and safety considerations forced TML to switch to using a cast-iron lining, after which the average drive rate fell to 25m per week. This level of progress was clearly unacceptable considering there was at least a further 3.5km of poor ground ahead of the TBM before conditions would improve. Measures urgently needed to be

Figure 3. Marine rail tunnel north: erecting segments (with modifications made to procedure). (Source: QA Photos)

taken to make the equipment on the TBMs more reliable and to try and stem the water ingress and reduce the volume of overbreak.

Major modifications were undertaken to the north TBM in October 1989, 14 weeks after first encountering problems due to the ground conditions (Figure 3). On the south TBM (which commenced its drive around the time the north machine was first running into difficulties), these same modifications were undertaken prior to the TBM encountering the adverse ground, thus avoiding the problems experienced in the north tunnel.

Because it was not thought it would be necessary to grout rings adjacent to the build area immediately, grouting of the crown segments was undertaken off a grouting platform located between 90 and 135m back from the ring-build area. Access to grout rings adjacent to the build area was impossible due to the presence of the protective canopy over the first few gantries and to the location of the upper segment conveyor, which delivered the upper segments along the crown of the tunnel to the ring-build area. Groundwater was therefore allowed to enter the ungrouted annulus behind the rings over the 90m length of tunnel between the build area and the grouting platform. This water then flowed under gravity behind the rings and eventually poured into the build area, severely hampering the miners struggling to build rings, and also causing major electrical problems to the segment erectors. The only way to solve this problem was fully to grout rings adjacent to the build area, thus stemming the water at source. Early grouting of rings was also essential for the purpose of stabilising rings which had been expanded against packing (movement of the packing could cause destressing and collapse of a ring).

Thus, major modifications were carried out primarily to allow early grouting of the rings, and involved the removal of the protective drip canopy, removal of the upper segment conveyor, and the installation of an upper grouting platform. This platform provided access to crown segments from the penultimate ring built to 30 rings back down the tunnel. The removal of the upper segment conveyor meant that all segments had to be transported to the build area via the lower segment conveyor. The potential maximum advance rate of the TBM was reduced by these modifications.

At the same time, a major exercise in waterproofing the electrical equipment was undertaken to improve TBM reliability. This work included the removal of programmable logic control (PLC) equipment and connections and replacement with control via direct wiring to equipment, replacement of standard cabling with armoured cabling, and improvements to the earthing of the 24V control circuit supplies to moving parts of equipment. After these modifications, which took about three weeks to complete, progress of the north TBM improved dramatically, averaging over 100m per week and peaking at 177m per week in the wet and blocky ground conditions. After passing out of the poor ground, production rates immediately increased to average over 200m per week, peaking at 276m per week prior to the further modifications described later.

The last 700m of the zone of poor ground was considered to be particularly adverse, judging from the pilot tunnel experience. It was therefore decided to improve the ground conditions ahead of the north and south TBMs by injecting grout through holes drilled from the marine service tunnel. The grouting was intended to give a 3m annulus of treated ground around the top half of each rail tunnel. This was achieved by drilling 13–21m-long injection holes in a fan-shaped pattern every 3m of the tunnel length. Grouting was undertaken using the *tube à manchette* method (see Chapter 2). The advantages of this method of grouting are that it allows the separation of drilling and grouting works, grouting in short stages each of 500mm in length, and the possibility of regrouting any stage of a hole. A low-viscosity, silicate-based grout was used, injected under a carefully controlled pressure just to induce hydrofracture. The preference for using this method of grouting had been established earlier in the project through undertaking full-scale grouting trials. Subsequent permeability tests on the trial sections showed this method of grouting to be the most effective in the chalk marl. The success of the grouting of the 700m of particularly adverse ground in the rail tunnels was demonstrated by the rate of progress of the TBMs through the grouted sections. In both cases, progress rates exceeded 200m per week in the treated ground.

As described, the first major modifications reduced the optimum drive rate of the TBMs. Because of the earlier delays, it was necessary to carry out further modifications to increase this optimum drive rate beyond the 276m per week maximum rate that had been achieved after the initial modifications. These phase 2 modifications were undertaken during a three-week stoppage to both the north and south TBMs in May and June 1990, respectively. The modifications involved the replacement of the unreliable lower segment conveyor with a magazine loader, similar to those used on the Howden TBMs. In addition, the double handling of upper segments by the two sets of segment erectors was obviated by removing the lower segment erectors, and modifying the upper erector beam so that the upper erectors could pick up segments directly from the magazine loader in the tunnel invert.

Figure 4. Movement of marine rail tunnel north TBM through UK crossover cavern. (Source: QA Photos)

These modifications had an immediate beneficial impact on progress, although drive rates were initially constrained by the passage of each TBM through the marine crossover (Figure 4) and by logistics problems (shortages of trains due to the five concurrent drives). However, once these constraints had been overcome (by trains being released to the marine rail tunnels from the completed land drives, for example) an average rate of 340m per week was achieved over the last seven months of each drive. The maximum rate achieved by any TBM on the entire project was 428.5m in one week by the marine rail tunnel south TBM in early 1991.

The north marine drive was completed on 5th May 1991, the TBM having completed 17.9km of tunnel from its starting point at Shakespeare Cliff. The 19km-long south drive was completed on 8th June 1991. The higher average rates achieved over the first half of the south drive can be attributed to learning from the experiences on the north drive. This resulted, *inter alia*, in

Figure 5. Breakthrough of French TBM 'Europa' into UK marine rail tunnel north, 22nd May 1991. (Source: QA Photos)

the phase 1 modifications being undertaken to the south TBM prior to entering the wet and blocky ground.

Whereas the marine service tunnel TBM was driven to one side of the tunnel for burial, the rail tunnel TBMs were driven down below the line of the tunnels. The short sections of downwards drive were supported temporarily by shotcrete and the breakthroughs were effected by the French TBMs driving over the top of the buried UK TBMs. The official breakthroughs of marine rail tunnels north and south occurred on 22nd May (Figure 5) and 28th June 1991, respectively. By this time, both UK drives were more or less back on the original programme. The breakthroughs actually occurred ahead of programme (three months ahead for the north drive and two months for the south) owing to the better than expected performance of the French TBMs.

The progress of the TBMs over the second half of each drive more than made up for their poor performance during the first half of the drives. However, this improvement was achieved at considerable extra cost. In addition to the direct costs of the TBM modifications, the following factors contributed to the increased costs of the drives:

- Initially, a TBM gang comprised 34 men, but owing to the switch to more labour-intensive grouting methods, among other factors, the size increased to 46–50 men after the modifications.
- Bonus rates for building rings were increased to encourage production.
- Additional supervisory staff were required.
- Additional rolling stock was required to service the high drive rates experienced in the latter halves of the drives, also involving additional labour and maintenance.
- Additional grouting was required to fill voids and stabilise rings. Ground treatment was used for 700m of each rail tunnel.

Despite the unanswered question regarding the foreseeability of the adverse ground and hence the suitability of the TBMs and lining for use there, the TML and Eurotunnel tunnelling teams based at Shakespeare Cliff did a magnificent job in rescuing a situation which at one stage appeared to jeopardise the entire project.

Performance in land rail tunnels

The land rail tunnel TBMs commenced their drives on 2nd August and 27th November 1989, respectively, the second after breakthrough of the land service tunnel. Because of the experience in the land service tunnel with the unstable ground towards the end of the drives, fingers and infill plates were fitted to both rail tunnel TBMs above the build area to provide protection to the miners and to limit the extent of any overbreak.

Unlike the marine rail tunnel TBMs, all the segments were delivered through the invert on the Howden TBMs. Also, because dry conditions were known to exist for the most part of the land drives a protective canopy was not installed on the trailing platform behind the land TBMs. Thus, when overbreak did occur, the rings could be fully grouted adjacent to the build area since there were no problems due to lack of access to crown segments, as experienced on the marine rail tunnel TBMs.

Again, unlike the marine TBMs, no reliability problems were experienced due to breakdowns of electrical equipment exposed to saline water ingress. Whether the equipment on the TBMs would have proved reliable in wet and saline conditions was never put to the test since, for the most part, the ground was dry.

The only problem experienced with TBM equipment was the strain put on the segment erectors owing to the very thick and heavy segments used, i.e. 540mm-thick segments weighing up to 8t each. Although the erectors had been designed to handle these heavy segments, strengthening and other modifications had to be undertaken to reduce wear and tear on the erectors.

As a result of the problems experienced as the land service tunnel TBM passed through the low cover section immediately prior to its arrival at the Holywell reception shaft, it was decided to construct a 20m-long reception chamber for each land rail tunnel TBM. These chambers were built using a road header and NATM support and were located adjacent to the TBM reception shafts. Treatment of the ground beyond the reception chambers was undertaken over a distance of 0.5km to reduce water ingress into the tunnel drives and to improve stability. The grouting was undertaken from the land service tunnel in a similar manner to that described for the treatment of the marine rail tunnels. Despite these measures, problems were still experienced with blocky ground of a similar nature to that experienced in the marine rail tunnels, i.e. deformation of the fingers, packing of overbreak, and so on.

Because of constraints imposed on the land drives by shortages of supply trains, due to the marine tunnels having priority over the less critical land tunnels, the full potential of the land rail tunnel TBMs was never achieved. The post-commissioning average drive rates of the north and south land rail tunnel TBMs were 178 and 194m per week, respectively, with a maximum weekly total of 322m being achieved by the south land rail tunnel. Breakthroughs into the TBM reception chambers were achieved by the north and south machines on 11th September and 20th November 1990, respectively.

The south land rail tunnel boring machine is on permanent display outside the Eurotunnel Exhibition Centre in Folkestone.

Construction logistics

TIM GREEN

While the technology involved in building the tunnels has been truly remarkable, the perhaps rather humdrum logistical support needed to achieve the work lay at the heart of the project. This chapter concerns the UK logistical services and covers the delivery of goods to site, the construction railway, the standard-gauge railway in construction mode, and various services which used the service tunnel after the construction railway was removed. The key role of the pithead site at sea level at Shakespeare Cliff is described.

Some statistics give an idea of the sheer size of the Channel Tunnel project and the importance of unfailing support services. In the five years the construction railway was in use the works trains covered sufficient train miles to have made five return trips to the moon; at peak they travelled enough miles to circumnavigate the globe weekly. Seventeen million tons of spoil were brought to the surface at Shakespeare Cliff; ten million tons of materials, largely concrete segments, were delivered to the working faces. For a high proportion of the construction period 1000 people were taken to work and brought from work underground every eight hours of 364 days each year. Over 200 diesel powered or electric locomotives and other prime movers hauling well over 1000 wagons, skips and trailers were used at various project stages. These vehicular services were of three main types: construction narrow-gauge railway; standard-gauge railway; and rubber-tyred vehicles for use in the service tunnel. These three main logistical services were dominant in turn in the order listed. All three were in use for a while in 1993.

CONSTRUCTION RAILWAY

The first and by far the most heavily used of these services was the narrow-gauge construction railway. Its principal roles were to bring spoil from the tunnels, carry in materials, act as mobile working platforms, move people to and from their work, and as the main form of escape in an emergency.

The track comprised 34kg/m steel rails at 900mm gauge held on steel sleepers by Pandrol clips. The steel sleepers in the service tunnel were screwed to precast reinforced concrete slabs which supported them and formed a level surface above the curved tunnel invert. In the rail tunnels the sleepers were attached to precast concrete blocks which were kept in place either side of the tunnel centre line by dowels into the curved tunnel invert. The concrete blocks were not continuous; and gaps between the blocks were filled with ballast. There was a track either side of the tunnel centre line. In both the service tunnel and the rail tunnels these methods of track support proved to be sound and durable. Some maintenance problems did arise in the service tunnel, particularly in the cast-iron ring sections where the track was supported on steel beams across the tunnel invert. At the surface some track was concreted in and did require some maintenance. Most of the track at the surface was more conventional sleepered track on ballast on newly made ground which required maintenance by the usual methods of packing and slewing to retain standards of line and level.

The two tracks in each tunnel had a facing and trailing crossover every 375m to facilitate switching between tracks and to enable a single track to be used to get past work sites on the adjacent track. At the foot of the main access adit the six pairs of track came to a complex switch and crossing layout which enabled trains to move to and between the landward and seaward tunnels. The pit bottom area was connected to the surface track layout at sea level (pithead) by a five-track rack and pinion railway at a gradient of 1 in 7. At the surface the construction railway gave access to two major workshops, a platform from which tunnel spoil and rubbish could be tipped, a battery charging station, a concrete batching plant, the segment stacks – which were also served by standard-gauge tracks connected to British Rail between Folkestone and Dover, by which route the segments were delivered from TML's concrete lining factory at the Isle of Grain in North Kent (see Chapter 5) – and various lay-down areas used as staging posts for the catholic range of materials needed underground. All switches and crossings were hand operated by levers.

Power to move the trains on the rack railway and the three seaward tunnels (for most of the time) was provided from a 550V rigid overhead catenary. The electric locomotives also had batteries for their movements at the surface, in the pit bottom and in the tunnels close to the tunnel boring machines (TBMs). Diesel locomotives were used in the land tunnels. Before the two marine rail tunnels broke through with the French it was possible, owing to the excellence of the temporary ventilation system, to remove the overhead catenary and change to an entirely diesel fleet of prime movers.

Locomotives and rolling stock

The locomotive and rolling stock fleet comprised some 200 locomotives and 1000-plus wagons of many different styles. The electric locomotives for heavy

Figure 1. Construction locomotive (diesel) in a rail tunnel. (Source: Mike Griggs, TML)

trains worked as tandem pairs, while single electric locomotives were also used. The diesel locomotives with pre-ignition clean-burn engines were either 150 or 200HP (Figure 1). Prime movement was also achieved for special duties such as working rigs by two powered winches on a flatbed with cables attached to the track fore and aft.

Manriding was by two-car and three-car diesel sets. The rolling stock fleet included 330 purpose-made side-tipping 14 tonne muck skips which discharged their loads at the pit bottom into below-track bunkers from which the spoil was moved by a complex of conveyors around the pit bottom and to the surface via an adit, which had been part of the 1974 works (see Chapter 5). Most of the rest of the rolling stock were flatbeds, 6m long, which were used to take materials from the surface to work sites underground. There were a number of specials to carry cable drums and a fleet of 50 concrete-carrying 6 and 9m^3 wagons known from their shape as 'concrete bullets' (Figure 2). These were used to convey wet prebatched concrete from the surface plant to work sites up to 25km away underground.

The fleet of locomotives and rolling stock was maintained on site by a team of people 350 strong at peak. The team made many modifications and improvements to both locomotives and rolling stock to overcome problems

Figure 2. Rack and pinion locomotive hauling 'concrete bullets' in adit A2. (Source: Mike Griggs, TML)

associated with being put to work continuously for such long periods. Availability overall achieved against a 24h/day, 365 day/year demand was of the order of 95%, the downtime being a mixture of modification, planned maintenance and out-of-course repairs.

Pithead

The pithead at the foot of Shakespeare Cliff lay alongside the Dover to Folkestone railway. It served two major purposes during the construction phase. The first related to tunnel spoil: some 17 million tons of chalk marl was disposed of there behind an enclosing sea wall – England is 34ha larger as a result of this activity. Its other major role was to provide space for maintenance facilities for the construction railway, field plant, etc. and to serve as a collecting point for materials to go underground via the rack and pinion railway (Figure 3). At later stages it was the staging post for disposal of redundant temporary materials from underground. Temporary drainage pipes, electric cables and ventilation trunking were amongst many other materials passing through.

The pithead site is now home for the cooling water chillers and a number of other permanent facilities, clustered close to the access adits at the eastern end of the site. The rest of this very large area is landscaped and grassed. It contains two freshwater lakes and, as intended, is already attracting wildlife.

Figure 3. Pithead with BR locomotive bringing in last load of concrete tunnel segments. (Source: Mike Griggs, TML)

Functions of construction railway

The construction railway was used for many tasks, but of greatest importance was keeping the TBMs hard at work. In the rail tunnels each 1.5m-long tunnel ring required one train trip to service its needs. The trip to the TBM needed five flatbeds to carry the nine concrete segments that made up each ring (Figure 4), lengths of railway track to extend the railway and materials to serve the grouting and other TBM operations. The same train when it reached the pit bottom would collect empty muck skips. The return journey consisted of loaded muck skips and empty flatbeds. In the smaller-diameter service tunnel one such train on its journey in and out enabled two rings to be built. Each marine rail tunnel TBM was served by eight skip trains and 11 sets of flatbeds which permitted the pit bottom to pithead and back cycle which the muck skips did not have to make.

The UK crossover excavation and lining was another major user of the construction railway for material delivery and spoil disposal. The hand excavation of cross-passages, equipment rooms and piston relief ducts were sources of spoil for disposal and were also material hungry. Many sites needed prebatched concrete frequently in a sequence of trains of 'bullets' for major pours of *in situ* concrete which could last for many hours. Untoward

Figure 4. Concrete tunnel segments being loaded on to a narrow-gauge construction flatbed prior to transit to the marine tunnels. (Source: Mike Griggs, TML)

delay of a long tube of wet concrete gave rise, from time to time, to a fearsome task of excavating hard concrete from the bullets.

After completion of the tunnels, much of the fixed equipment – pipes, cable, brackets etc – was installed from construction railway flatbeds which conveyed materials to site and by much design ingenuity acted as powered specialist tools for placing the equipment on the tunnel intrados (Figure 5).

Moving people was a major task. Work patterns changed frequently such that published timetables, for the three 8h and two 12h shift changes each day, rarely lasted more than two weeks before some fine-tuning was required. Most manriding was by the purpose-built three- and two-car diesel sets. On some muck skip trains a manriding car was included to increase the availability and frequency of manriding between shift changes.

Many visitors were taken underground. Employees whose duties did not include visiting underground sites had the opportunity, on Saturday mornings,

Figure 5. Narrow-gauge rig in operation in a rail tunnel during 21kV power cable laying. (Source: Mike Griggs, TML)

to take a manriding car to one of the TBMs and the UK crossover cavern. A two-car manriding set was adapted for the purpose by fitting glazed sections in the roof and side to maximise the view on the journey. The source of the idea gave this set its name – the Disneymobile. VIP visits were also frequent. A number of members of the royal family had visits underground: the Duke of Edinburgh made an historic trip on 3rd April 1992. The UK and French construction railways were connected in a rail tunnel for the occasion and the Duke made the first crossing by rail from the UK to France in the Disneymobile. For these VIP visits the two-car Disneymobile was split and another very special manriding car was marshalled between. Its first use was in early December 1990 for the breakthrough in the service tunnel. Its name – the Maggiemobile – turned out to be a misnomer due to a political change but it has nevertheless survived. Padded seats and carpets on both the floor and sides gave this vehicle some degree of comfort above the normal for such vehicles.

That there was an organisation (Underground Operations) that ran the railway for the very many and wide range of customers who needed logistical support was probably a world first. Normally tunnellers control and run their own railway. The Channel Tunnel project was so large that unification of the logistical management made sense. Despite the unified control of the railway, which had common methods of operating, professional drivers, experienced controllers, etc., each working team had a dedicated team of drivers. The competition between teams, particularly of tunnellers, was fierce.

There was no signalling system other than a traffic light system at the pit bottom and on the rack railway. Train movements in each tunnel were controlled by one operative, the operations board controller (OBC), who was in two-way radio contact with all the train drivers. He had a mimic board which showed the track layout and to which he could attach magnets depicting each train and manriding car. Sections of track, between crossovers, which were occupied by trains serving work sites were shown in a different colour. The OBC had to work tunnel traffic past such sites on the single line.

The busiest tunnel was frequently the service tunnel. Typically up to a half of the tunnel length could be subject to single-line working to some 20 different work sites. The OBC had usually to control 35 or 40 inward and a similar number of outward journeys through the complete layout on his 8h shift. The OBCs were key people who learned skills comparable with those who control aircraft movements.

STANDARD-GAUGE RAILWAY

As standard-gauge track was progressively laid from the Folkestone Terminal towards the UK border so the two-track construction railway was removed. But the fixed equipment installation was not then complete. At this time a fleet of 25 standard-gauge locomotives, again with flatbeds with all sorts of ingenious rigs and tools, were brought into use. Entry to the tunnel complex was from the Folkestone Terminal where a pithead had to be created on temporary track to the north side of the site. For most of the time that this arrangement was in place each rail tunnel track was effectively a long siding. Trains which went in forward had to reverse out (unlike the construction railway where the locomotive could use two crossovers to run round the train and thus haul the train for both inward and outward journeys). On the standard gauge a second man rode a caboose on the rear of the train and when reversing used a hard-wired telephone or a back-to-back radio to keep the driver informed. The caboose man also had a horn to warn pedestrians and an emergency air-dumping brake valve.

At about this time, the project changed to 12h shifts, so all 25 trains had to come out and go back in the course of 12h. Control was by radio. On the standard gauge the controllers were called rail movement controllers (RMCs). The initial test runs with shuttle rolling stock was controlled by the RMC until the permanent signalling system was commissioned. For the

Figure 6. Standard-gauge BR diesel multiple unit in the UK crossover. (Source: Mike Griggs, TML)

first time links with France had to be introduced as international movements became commonplace.

Standard-gauge BR diesel multiple units – two complete three-car sets – were used as the first train in and last train out of the rail tunnels at shift changes (Figure 6).

On removal of the construction railway tracks from the service tunnel, a smooth concrete screed was laid, upon which the permanent service tunnel transport system (STTS) vehicles would run. The rail tunnel, with a single track, had logistical value but little flexibility. It was for this reason that nine diesel-powered, rubber-tyred articulated manriders with a total of 34 seats each were used in the service tunnel. Working rigs in the service tunnel on the concrete screed were mounted on braked, rubber-tyred trailers hauled by one of two sizes of tractor.

As the opening of the Channel Tunnel approached, the STTS vehicles also ran on the service tunnel screed. The powered vehicle movements, all on rubber tyres, in the service tunnel were controlled as before by radio by the OBC. Work sites created single-line working sections and control was exercised as if it were still a railway.

For the last 12 months or so of the project there were many people who had the need to visit 10 or 15 sites in one trip over as many kilometres. Tying up a dedicated 34-seater manrider would have been profligate; asking an engineer to wait for the 40 minute interval service between shift changes to move on a couple of kilometres would have been a waste of time. A fleet of 125 mountain bicycles was introduced. They were not radio controlled but had very visible markers! But is it really possible to discipline a supervisor who, contrary to procedures which required a cyclist to stop when a powered vehicle was about, was caught riding up the 18km long 1-in-90 gradient and actually overtook a 34-seater manrider doing 25km/h!

SUCCESSFUL LOGISTICAL SUPPORT

At its peak, the Channel Tunnel temporary railway was the third largest in the UK. While many users complained about the quantity of support available it is significant that most of the tunnelling records were broken when the logistical chain was at its longest.

Those who drove the tunnels were justly proud of their records. Perhaps the most telling of these was the 8h shift record in the rail tunnels: 20 rings, each 1.5m long – a 7.8m diameter tunnel advance of 30m. The logistical support on such a shift involved not only the 20 segment trains moving in and the 20 spoil trains moving out but three or four manrider journeys, at shift change, both in and out, and possibly another five or six trains in and out serving cross-passages and piston relief ducts and a similar number serving fixed equipment installation sites on the route into the main tunnel face.

The Channel Tunnel project was of world class. The logistical support to all the types of work played no small part in its success.

III
Railway/Tunnel Services

8

Radio links

KAREL DE JAEGER-PONNET

Passengers traversing the 50km of the Channel Tunnel will no doubt wish to feel confident that the train operator is always in contact with dry land at either end. The means by which such contact is ensured involves both hard and soft systems.

All the control and communications networks depend on three fibre-optic cables, each of which is carried through one of the tunnel bores. Speech and data are transmitted digitally along these cables, which are of very high capacity. The transmission network is configured so as to ensure essential services even if any two of the three cables fail. In addition to speech communications, these fibre-optic cables carry data for rail traffic management and all other electrical and mechanical plant installed in the tunnels.

Tied in to the hard-wired communications networks are a number of independent radio systems:

- **Concession radio (CR)** provides a general means of communication for mobile vehicles and personnel working throughout the limits of Eurotunnel's Concession covering the tunnels, terminals, and the two sites above the coastal shafts.
- **Track-to-train radio (TTR)** provides a secure speech and data channel between the locomotive cab and the railway control centre.
- **Shuttle internal radio (SIR)** allows the crew of a shuttle to talk to each other, and to passengers, over their car radios.

In addition, there are three radio systems that are only used above ground in the terminal areas. These cover paging, the remote control of shunting locomotives, and loading and unloading operations on the freight shuttle platforms.

Emergency services, notably fire, ambulance and police, have their own tactical radios which would not work satisfactorily within the tunnels owing to attenuation. Arrangements have therefore been made to extend the range of their hand-portables using tactical repeaters on two service tunnel transport system (STTS) vehicles, and antennas radiating simultaneously in the three tunnels. These repeaters can also be linked to telephone outlets in the tunnels.

This tactical radio system effectively forms a fourth radio network for use in emergency situations. In addition, the emergency services use the CR handportables for calls throughout the Concession and to the outside world, the CR system ensuring that such calls receive priority.

Motorola Storno supplied these Channel Tunnel radio systems. Special features include the ability of equipment to withstand the hot, moist and saline environment within the tunnels, and compatibility with regulations and standards applying in both Britain and France. The TTR system, which applies to through trains as well as shuttles, is compatible with the BR train radio specification, but not with that of SNCF.

The three in-tunnel radio systems have to be continuously available, which means that no single failure must lead to the loss of the complete system, and the effect of any failure is minimised.

CONCESSION RADIO

The CR is there mainly to allow people working in the terminals and tunnels to talk to each other, effectively extending the telephone network to mobile users. It can also handle limited data transmission. The initial number of mobile radios, either personal or mounted in the shuttles, locomotives of the national railways, and the STTS vehicles, was around 800.

Voice calls can be made to an individual or to a specified group. The latter can take the form of conference calls, or they can be broadcast to the group by one individual.

Data can be transmitted either in the form of a status report (one of 30 different states), or a simple text message of up to 200 characters; up to 10 such messages can be stored by a mobile radio for recall by the user on demand.

Any CR user can be connected to the normal Eurotunnel telephone network, but regular users working at fixed locations can have telephones permanently wired in. This phone can be linked to a VDU workstation, allowing incoming calls to be monitored and text messages to be composed on a keyboard for transmission to an individual or group.

CR is used for managing the flow of both road and rail traffic within the terminals, and it allows the shuttle train drivers and chefs de train to communicate with the railway control centre as well as other areas such as the maintenance depot. So far as train drivers are concerned, it backs up the TTR which is intended to provide a secure speech and message channel for authorising the safe movement of trains under normal, emergency or failure conditions.

There are 40 CR base stations, 35 of which are located in technical rooms between the tunnels at 1.5km intervals. Radiating cables of the low-smoke, nohalogen, fire-retardant type extend for 750m in both directions along each of the three tunnels from the base stations. Each of the terminals and coastal sites has its own base station, the Calais Terminal requiring two because of its size. Radiating cables are installed in the cut-and-cover tunnel carrying shuttles around the loop at the Folkestone Terminal.

Remote control of the 40 base stations, the allocation of channels to users, management of user databases, and channel switching are exercised by a master switch controller (MSC) at Folkestone, and three trunking switch controllers (TSCs) covering respectively the two terminals and the tunnels. The MSC is used only when it is necessary to connect two radios in different TSC areas. Each TSC acts as a backup for one of the other two, except for the Sangatte shaft base station because of frequency allocation restrictions.

If the MSC fails, calls between TSC areas become impossible, and if two TSCs fail simultaneously, users can only talk to radios covered by the same base station. There is substantial redundancy in the links between base stations and TSCs to ensure that alternatives are generally available.

TRACK-TO-TRAIN RADIO

Locomotives, power cars and on-track maintenance vehicles that can run on Eurotunnel tracks are fitted with TTR. This links the train driver directly to the railway control centre, normally at Folkestone but exceptionally at Calais.

The primary function of TTR is to transmit voice and data messages that affect the movement of trains under all circumstances, but particularly when the normal signalling may be degraded or inoperative. The system must therefore ensure that the message is received by the correct train, or in simple terms, the controller must be in no doubt as to which train driver he is addressing. Any other driver can initiate an emergency call, although it is up to the controller to break off his existing conversation and answer it.

For operational purposes, Eurotunnel's tracks are divided into four areas comprising the two terminals and the two rail tunnels. Only one conversation can take place at a given time within each area, though it can be a general call to all trains.

Despite the specialised purpose of the TTR link, the controller does have the option of connecting a train driver to the telephone network. Alternatively, a shuttle train driver could use the CR for this purpose.

The TTR closely follows BR's 1845 specification, with which it is fully compatible. One feature lacking is the ability of the controller to talk directly to passengers in shuttles through the train's public address system; this can only be done by the driver or chef de train.

A call from a train cannot be transferred by a controller to a supervisor, although a three-way conference call is possible. This ensures that the supervisor does not give instructions to a driver without the controller responsible for that part of the railway being aware of it.

Each train driver sees a three-line display of data messages. The first line normally includes the radio channel, the area code, and the train running number. The other two lines can display brief messages of up to 16 characters per line in English or French, as the driver chooses.

Within each rail tunnel, main and standby base stations are provided to enhance reliability. TTR control systems are duplicated for the same reason, one being located at the control centre in each shuttle terminal. On the shuttle locomotives, the TTR mobile radios are completely duplicated, with main and standby panels on the driver's console and a third panel in the rear cab.

As with the CR, base stations are installed in equipment rooms at 1.5km intervals, with 750m-long radiating cables, in each rail tunnel. However, they are halfway between the CR base stations so that a fire in one equipment room will not put both CR and TTR systems out of action. Each terminal has its own base station, and as with the CR, a radiating cable covers the underground section of the loop at Folkestone.

So that radio coverage corresponds to the four signalling control areas, the terminal radio areas extend into the rail tunnels (using radiating cables) for 350m from the French portal. At Folkestone, coverage extends 1.5km from the portal, embracing the crossovers in cut-and-cover tunnel between Castle Hill and Holywell Coombe.

SHUTTLE INTERNAL RADIO

Each passenger shuttle has its own internal radio system operating simultaneously in the UHF and VHF bands. One UHF and two VHF transmitters are installed in the locomotive at each end of the train. They are all connected to radiating cables installed within each deck of each wagon.

The SIR performs two distinct functions. First, it allows the driver and train captain to talk to other members of the train crew, who have personal radios; this is an open channel heard by all users. Secondly, messages recorded in English and French can be broadcast simultaneously to the occupants of all cars who have tuned in their radios to the advertised frequency.

There are also wired communications throughout the shuttle: two telephone systems (routine and emergency), passenger alarm and public address system. When the public address system is used, normally by the *chef de train*, it automatically overrides recorded music or messages being transmitted to car radios.

Portable UHF radios carried by the shuttle crew can be used either in CR or SIR mode. They will normally use the SIR mode during loading and unloading passenger shuttles when out on the platform or during the journey. The only time they will switch to the CR mode is when evacuation of a shuttle in the Tunnel is necessary due to an emergency.

9

Power system

JOHN FINN

The Channel Tunnel railway relies absolutely on electricity to operate its many systems. It is of prime importance that the supply should not fail.

The power system is itself unusual in that it normally has two sources of supply, one from the French system and one from the British system, which must never be paralleled. The total load of the system is shared approximately 50/50 between the two sources of supply.

The nature of the load can also be split into two categories: traction load and auxiliary load (Figure 1). The loads on each side are made up of approximately 120MVA of traction and 45MVA of auxiliary load for the year 2003 developing to 180MVA of traction and 45MVA of auxiliary load at the ultimate capacity of the fixed-link transportation system.

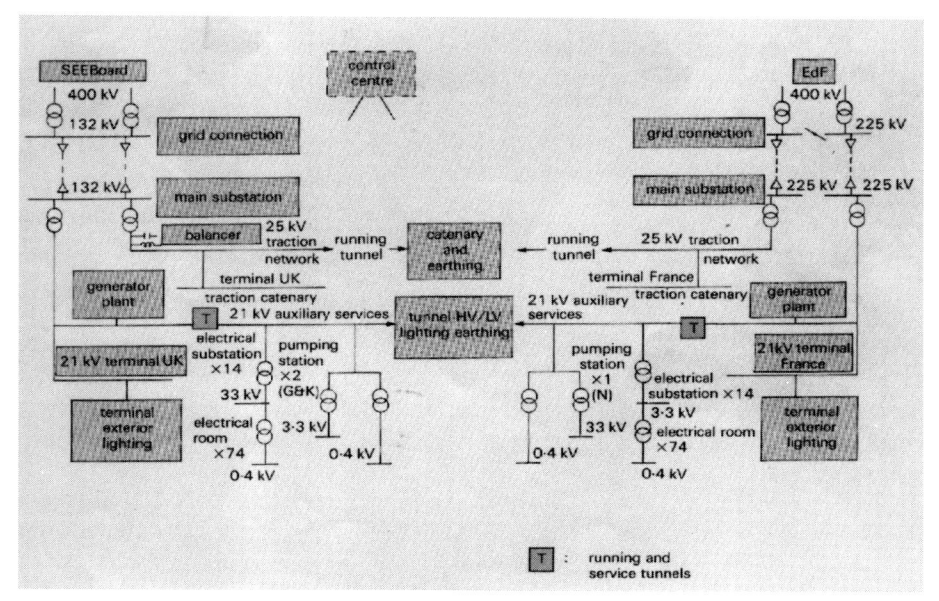

Figure 1. Overall diagram of power supplies

Whereas the power system is designed to ensure that the two sources of supply are never paralleled, the earthing system is connected solidly throughout. As with all 25kV AC overhead traction systems, the return current is through earth and care is taken to ensure that the steady-state touch and step potentials do not exceed 25V. For fault conditions, the relationship of touch voltage with regard to fault clearance times has been based on the most recent work published in IEC 479-1 (1984).

To develop the design of the power system and also to subcontract the work in logical, manageable packages, the system was divided into a number of primary subsystems:

- grid connections
- main intake substations
- tunnel high-voltage distribution system
- tunnel low-voltage distribution system and lighting
- shafts and standby generators
- terminal distribution systems
- exterior lighting of the terminals.

These primary subsystems exist independently in France and the UK, except for those dealing with the tunnels themselves, where the system has to be completely integrated throughout.

GRID CONNECTIONS

UK side

The load has a large traction content with its inherent unbalance and high harmonic content. Consequently, it is essential that the connection is made to a strong network to reduce the effect of these disturbances on other consumers. This necessitated a connection at 400kV; the nearest existing substation was at Sellindge approximately 14km to the west of the Folkestone Terminal.

The grid connection on the UK side consists effectively of duplicate connections each rated at 240MVA. These connections are effected in practice by extending the existing 400kV double busbar SF_6 GIS substation at Sellindge to accommodate an additional 400kV switchgear bay at each end of the busbars. These bays are connected to 400/132kV autotransformers. The two transformers are connected by 132kV oil-filled cable into a single busbar 132kV SF_6 switchboard housed in an extension of the existing 132kV switchroom and making use of the same cranage facilities. This increases the availability of the supply by permitting the simultaneous outage of either transformer with either of the main cable connections to Folkestone.

The connection to Folkestone is by two 132kV 1000mm² oil-filled cables each capable of 240MVA continuous rating under summer ambient conditions. The circuits follow the same basic routes but are sufficiently separated to prevent a mechanical digger from causing damage to both circuits simultaneously.

Even with the connection from 400kV the electricity supply authority imposes strict restrictions on the level of negative phase sequence (NPS) voltage and harmonics which are acceptable at the point of common connection. At Sellindge the NPS restriction has been set at 0.25% and levels varying between 0.15 and 0.4 set for harmonics from the third to the 25th. The fault level at Sellindge is normally in the region of 20GVA; however, under minimum fault level conditions, it may fall as low as 6GVA. Hence the maximum level of unbalance allowed at Sellindge is 0.25% on 6GVA, i.e. 15MVA. To solve this problem a load balancer is required on the UK side and this is located in the main high-voltage substation.

French side

The same principle of connection to a strong network applies on the French side and the connection is made to the 400kV substation at Les Mandarins. This part of the French network is particularly strong, being well supported by nuclear plant, thus giving a consistently high fault level. The maximum fault level is of the order of 20.5GVA and the minimum, which is more significant to the tunnel system, is 11.7GVA.

The French supply authority, Electricité de France (EdF), also allows NPS voltages of 1% and harmonic voltage levels of 0.6% for even harmonics and 1% for odd harmonics subject to an overall distortion level of 1.6%. This allows an unbalance of 117MVA at the point of common connection (1% of 11.7GVA) and thus removes the need for any load balancing equipment to meet this need. However, the impedance of the 400/225kV transformers reduces the minimum fault level on the main intake substation busbars to 2030MVA. As the modern design of motors only allows for 1% of NPS voltage continuously, this would effectively limit the unbalance acceptable to 20.3MVA.

As the distance between Les Mandarins and the Calais Terminal is short, approximately 2.5km, it was decided to provide a separate connection for the auxiliary system such that the EdF point of common coupling is the same as the point of common coupling of the traction and auxiliary supplies. Consequently, the French grid connection consists of three circuits, one normally associated with the traction power and one with the auxiliary power and the third (emergency bar) switchable between the two by automatic switching devices.

The grid connection on the French side comprises two 400kV circuit breakers connected into the 400kV mesh at Les Mandarins, each one feeding a 400/225kV 300MVA autotransformer. Normally, transformer 1 is connected to the 225kV traction busbar while transformer 2 is connected to the emergency busbar. The emergency busbar is normally connected to the auxiliaries busbar via a bus coupler circuit breaker. The circuits from stations Echingen and Les Attaques are also connected to the auxiliaries busbar. This makes good use of the equipment as it enables EdF to support the 225kV system using the supergrid transformer when both transformers are available while giving security to the auxiliary supplies from Echingen and Les Attaques when one of the supergrid transformers is out of service. There are three 225kV 630mm^2 XLPE cable circuits each rated at 210MVA summer rating connecting the associated bars at Les Mandarins and the Calais Terminal.

MAIN HIGH-VOLTAGE INTAKE SUBSTATIONS

The main substations take in power from the grid connections and transform this down to the required voltages for utilisation, namely 25kV single phase to earth for traction and 21kV three phase for the auxiliary system.

French side

To ensure that the traction and auxiliary systems are integrated and to have the flexible switching required, the main subsystem also has three busbars: traction, auxiliary and emergency, mirroring the system at Les Mandarins. The traction system uses 25kV single phase to earth. The terminal is fed from the Y phase while the rail tunnel south uses R phase and the rail tunnel north B phase.

The feed for each 25kV phase is obtained from 225/27.5kV, 60/75MVA ONAN/ONAF single phase transformers. There are three transformers connected to dedicated phases R-Y, Y-B, B-R, plus a further spare transformer which can be selected to any pair of phases on the HV side. These traction transformers have an impedance of 19.8% on ONAF rating to limit the maximum fault level on the 25kV system to 12kA, which is the maximum acceptable to the locomotive. The voltage on the traction busbars is allowed to vary between 27.5kV and 25kV depending on the traction load. All the 225 and 25kV equipment is conventional outdoor equipment.

The auxiliary supply system is secured by two 225/22.5kV auxiliary transformers rated at 30/40/45MVA ONAN/ONAF/OFAF, having impedance of 17.5% on rating to limit the fault level to 12.5kA with two transformers in parallel. The neutrals of the transformers are each earthed through 500A neutral earthing resistors, thus limiting the 21kV earth fault current to 1000A with both transformers in parallel.

The transformers are connected to an indoor 21kV single busbar switchboard using SF_6 switchgear. This switchboard is divided into five sections located in three separate rooms segregated by fire walls; the two transformers feed to the end sections with the outgoing circuits being spread across all sections. Provision has been made for the addition of a third transformer in the future should this be considered necessary.

UK side

The UK main substation performs the same function as the French substation, but the incoming voltage is 132kV. The 132kV system is a double busbar with a section/coupler arrangement in the middle between the traction half and auxiliary half of the substation. The normal arrangement is with the system run solid and both grid connection circuits connected. The auxiliary system feeding arrangement is exactly the same for the French side, except that the primary voltage is 132kV instead of 225kV. The traction transformer arrangement was also originally designed to be the same; however, the interaction of the balancer necessitated a completely revised design.

To understand the reasons for the differences in the main substation designs it is necessary to look in some detail at the problem of unbalance caused by the traction load. Initially the terminal is fed from the B-phase with rail tunnel north fed from R-phase and rail tunnel south from the Y-phase. The load on each phase varies according to how many trains are on that section of catenary and whether the trains are driving, coasting or braking. Consequently, it is a continuously fluctuating load on each phase with continuously changing patterns of unbalance. To present a balanced load to the 132kV

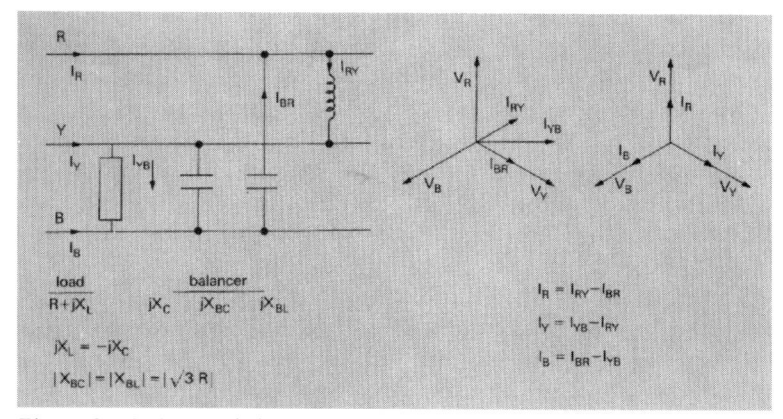

Figure 2. Steinmetz balancing principle for delta-connected load

system without using real power, a load balancer using the Steinmetz principle is employed.

Consider a single phase load connected between two phases of a three phase system (Figure 2). This unbalanced load can be balanced by power factor correcting the load and adding a capacitive load equal in magnitude to $P/\sqrt{3}$. Between two other phases and an inductive load equal in magnitude to $P/\sqrt{3}$ between the other two phases. It can be seen from the vector diagram that the resulting line currents drawn from the system are then balanced and in phase with the voltage vectors, i.e. it appears to the system as a balanced unity-power-factor star-connected load. Any combination of loads on the three phases can be considered by superposing the solutions obtained assuming each phase-to-phase load as a single phase load.

This explains simply how any fixed set of unbalanced loads can be balanced but, as explained, the traction system is generating a continuously changing pattern of unbalance. This is solved by connecting to each phase a fixed capacitive bank which can be backed off by a thyristor-controlled reactor (TCR). A sophisticated control system monitors the instantaneous loads and calculates the amount of capacitance or reactance required in each phase and sets the firing angle of the TCR to achieve the required effect.

The traction load, although connected phase-to-earth, would appear as a phase-to-phase load at the high voltage side of the transformers. However, to avoid the use of costly and space-consuming transformers the balancer is connected at 25kV in parallel with the traction load. This method of connection of the balancer effectively eliminates the negative phase sequence voltages and produces balanced currents at 132kV, but the side effect is that it doubles the zero-sequence current flow in the 25kV secondary. The effect of this zero sequence current flowing through the transformer impedance is to cause highly unbalanced voltages on the 25kV side.

This phenomenon is more easily understood by reference back to the simple case of balancing a single-phase load. It can be seen that the resultant three phase loads to create balanced 132kV currents are one phase resistive load, one phase capacitive load and one phase inductive load. Consequently, the resistive load basically causes a phase shift through the transformer impedance with little change in magnitude. However, the capacitive load causes significant voltage rise and the inductive load creates a significant voltage drop.

This voltage unbalance would cause problems for the operation of the 25kV system. The 25kV busbars need to be maintained within a voltage range of 25kV minimum, to ensure adequate voltage at the pantograph of the trains at the most distant point in the centre of the tunnel, to 27.5kV maximum, to prevent damage to traction equipment from overvoltage.

The problem arises from the voltage changes caused by zero sequence current flow. If a transformer with low zero sequence impedance is used this effect can be reduced. The use of an interconnected star or zigzag winding was chosen (similar to that used in earthing transformers). The connection of single-phase transformers in zigzag together with a switchable spare is not a viable solution. Consequently, the decision was made to use three three-phase star/zigzag transformers, each 50% rated.

There are two special requirements for the specifications of these transformers. First, although low zero-sequence impedance is required the positive and negative sequence impedances must remain high to limit the fault level to 12kA. Also, although the line currents are balanced on the 132kV side, the currents flowing in the low voltage windings are unbalanced and can exceed the normal maximum load current that would occur on a normal balanced three-phase system. The result is that the balancer eliminates the negative sequence effect while the zigzag connection minimises the effect of the zero sequence current, as seen on the 25kV system.

The traction equipment generates harmonics and the thyristor required for the balancer aggravates the situation by adding further harmonics. It is therefore necessary to provide filters to keep the harmonic distortion within the tight limits specified by the supply authority. These filters are incorporated into the capacitor banks required for the balancer by providing a mixture of damped and fully tuned filter arms to make the best use of the available capacitance (Figure 3).

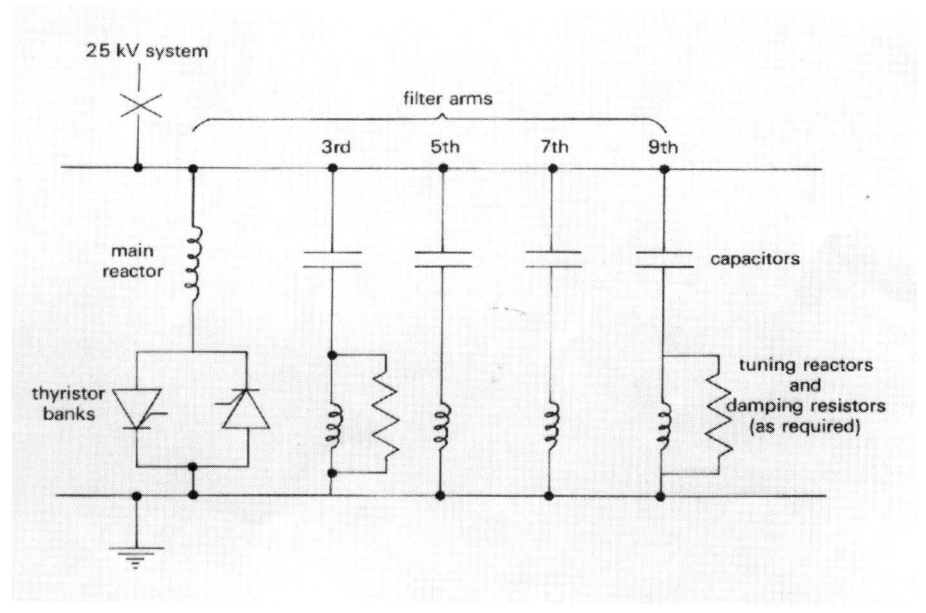

Figure 3. Diagram of one phase of balancer

TUNNEL AUXILIARY SYSTEM

The special requirements of the tunnel auxiliary system may be summarised as follows. First, although it is an integrated system, the tunnel auxiliary system will normally be fed from both ends, but under no circumstances are these two infeeds to be allowed to be paralleled. Secondly, the load to be supplied by this system splits into three basic parts:

- the largest part of the normal load which has a high motor content is concentrated at the two shafts at Shakespeare Cliff and Sangatte;
- the pumping stations which as a result of the tunnel profile are located at the low points of the Tunnel;
- the distributed small loads, lighting and socket supplies which require a number of substations and electrical rooms spaced at regular intervals through the Tunnel.

Thirdly, the system needs to provide security of supply to allow planned and unplanned outages. The system has also to have the ability to be rapidly reconfigured to allow for the total loss of supply from either the UK or France or both simultaneously.

From economic considerations 21kV was chosen as the normal voltage with an operating tolerances of ±5% giving typically 22kV on the busbars at each main substation with no less than 19.95kV at any point on the primary voltage system.

The 21kV system consists of two feeders to the pumping stations, one of which is normally fed from the UK over its full length and one of which is normally fed from France (Figure 4). This has the advantage that one circuit is maintained live to all essential loads, e.g. ventilation, firefighting and pumping stations, even under loss of one grid system.

This is supplemented by two cables feeding the distributed loads via a system of 14 electrical substations. Substations 1–7 are normally fed from the UK and 8–14 from France, with the circuits run open on bus section breakers at ES7.

These four cables provide the capability of dealing with all the priority loads. However, at Shakespeare Cliff and Sangatte, the nonpriority load of the cooling is connected into the pumping station feeder fed from the home side. To support this load and to improve security and voltage drop conditions, two additional 21kV cable circuits are run between the Folkestone Terminal and Shakespeare Cliff no. 1 board and between the Calais Terminal and Sangatte no. 2 board.

To ensure the necessary security of supply and improve operational flexibility, interconnectors are provided between the pumping station circuits and the electrical substation circuits at both shaft locations and at the pumping stations. In the event of the loss of a grid the system has load shedding applied

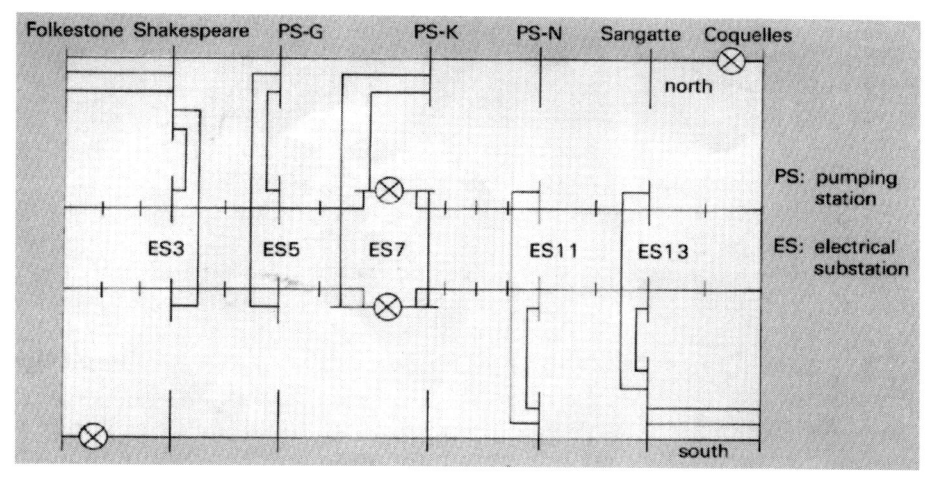

Figure 4. Tunnel 21kV distribution system

to reduce the load to that suitable for the downgraded conditions and the system is reconfigured to be run solid with all four cables in parallel to reduce voltage-drop problems under this onerous condition.

At those electrical substations not associated with an interconnector, ring main units consisting of two disconnectors and a fuse switch are used. At electrical substations where interconnection takes place circuit breakers are provided to sectionalise the 21kV electrical substation cable feeders.

To ensure fast, discriminated fault clearance, the prime feeders and interconnectors are fitted with pilot wire protection. The system is also equipped with overcurrent and earth fault, multiple in some locations to allow for the changing requirements under normal, downgraded and standby generator operation.

TUNNEL LOW-VOLTAGE DISTRIBUTION AND LIGHTING

The low-voltage system is required to distribute power from the electrical substations to the many and various auxiliary loads distributed throughout the tunnel such as:

- lighting
- piston relief duct dampers
- cooling valve actuators
- firefighting valve actuators
- catenary isolators
- cross-passage door electro/hydraulic units

- control and communication gear and switchgear battery chargers
- socket outlets.

These items all take power at 400V and originally it was intended to feed these from the electrical substation locations at 400V. However, with the long distances between substations varying between 2.8 and 5.1km this was not considered to be economically viable, and an intermediate voltage of 3.3kV was introduced.

Each electrical substation is divided into two parts, north and south, fed by separate 21kV cables (Figure 5). Each part has a step-down transformer feeding a main 3.3kV board. This board feeds onto radial 3.3kV cables, one feeding electrical rooms to the east, one feeding electrical rooms to the west and one feeding the electrical room adjacent to the substation. If a cable is faulted or taken out of service between two electrical rooms, the electrical rooms downstream of the faulted cable can be restored by linking onto the end of the radial feed from the adjacent electrical substation. Each electrical room, except the adjacent ones, houses a 3.3kV ring main unit feeding a 3.3/0.419kV transformer which supplies a 400V distribution board. The rooms are spaced at approximately 750m intervals. The 400V boards at each room are intercon-

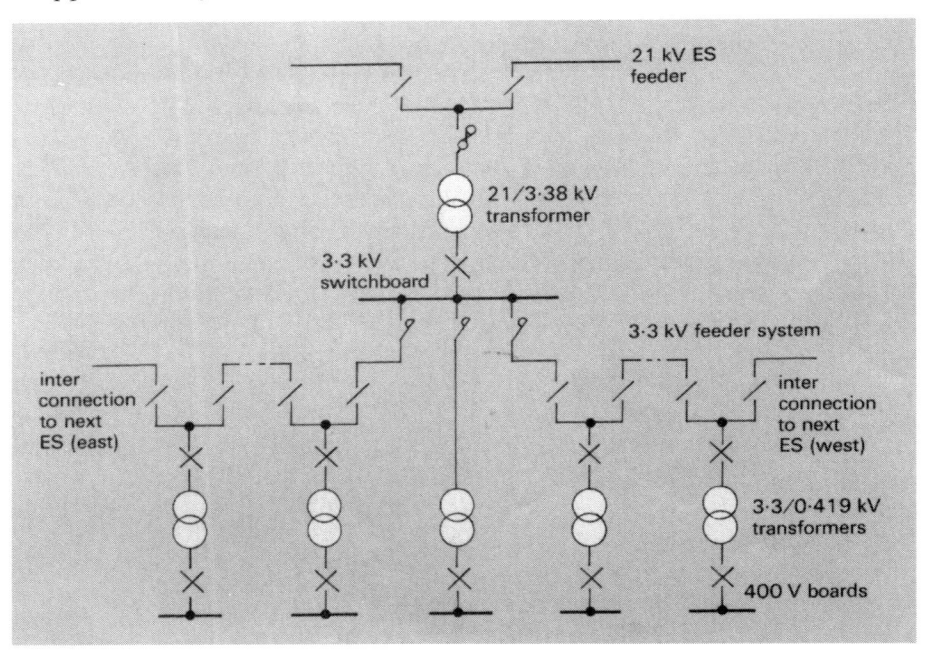

Figure 5. Typical 3.3kV radial distribution

nected by a 400V interconnector equipped with an automatic changeover with a three-out-of-four interlock scheme.

The lighting system (Figure 6) uses an insulated earth principle with continuous monitoring of phases and neutral to earth by a DC monitoring system. To create this separate insulated earth system 0.4/0.4kV isolating transformers are required at each electrical room. This insulated earth system enables the lighting circuits to remain in service even with an earth fault on one cable line, but gives an alarm so that maintenance personnel can investigate the problem.

The lighting is a critical safety item under tunnel evacuation requirements and this was the subject of much discussion before the final solution was adopted. The lighting of each 750m section of tunnel is covered by two overlapping circuits. One circuit is fed from the electrical room to the east of the section and the other from the electrical rooms the other side of the service tunnel and to the west of the section. Each circuit feeds every alternate lamp so that loss of one circuit will leave every alternate luminaire lit.

One of these two circuits for each section of tunnel uses special fire-resistant cabling to BS6387 class CWZ giving a 3h performance under 950°C fire conditions in the presence of water and with mechanical shock. This is to secure half the lighting for the remainder of a section in the event of a localised fire. The solution was used in preference to battery-backed fittings because the maintenance required for the batteries was considered impractical under tunnel operating conditions. Battery-backed fittings have been retained for the information markers indicating the locations of the cross-passages.

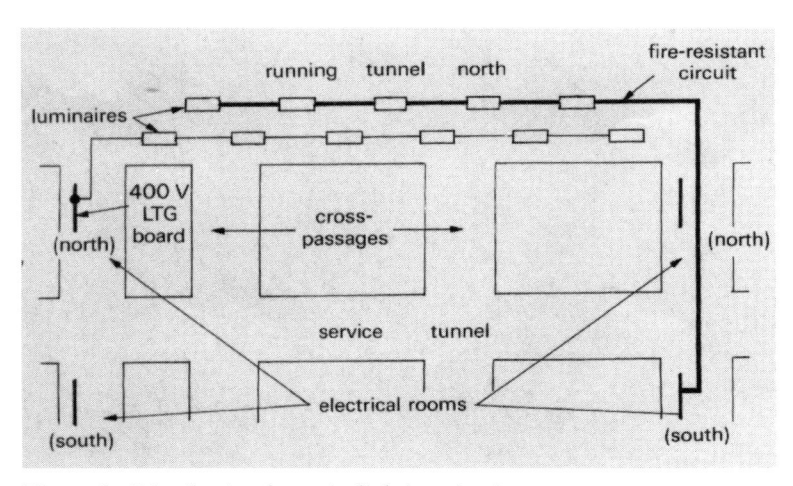

Figure 6. Distribution for main lighting circuits

The method of overlapping with one fire-resistant circuit is also employed in the service tunnel. The lights are switched so that initiation of the lighting for one section brings on the lights in the adjacent section both to the east and to the west thus illuminating approximately 2250m of tunnel. Initiating one section of rail tunnel lighting illuminates 2250m of service tunnel lighting and all associated cross-passages, but leaves the opposite rail tunnel in darkness. Initiation of the lighting can be either remotely from the control centre (the normal method) or from illuminated push-buttons located every 75m along the walkway inside the running tunnels.

The luminaires have had to be specially developed to provide aerodynamic shaping and reduce the drag in the rail tunnels and also to be able to achieve the required IP 65 rating under dynamic pressure conditions. They provide a 20lux lighting level with 0.5 uniformity after allowing a 0.75 downgrading factor for dirt and ageing. This was achieved by lighting mounted 4.1m above the walkway with 8m spacing using 18W fittings.

See also Chapter 17, on lighting.

SPECIAL CONSIDERATIONS FOR TUNNEL EQUIPMENT

The physical locations of the switchboards and the cable feeders have been carefully selected to avoid a single incident affecting more than one of the duplicate supply arrangements. At each pumping station and electrical substation, the duplicate switchboards (north and south) are located in separate rooms either side of the service tunnel, i.e. one room located between the service tunnel and rail tunnel south (Table 1).

Table 1. Segregation of cabling and rooms

Rail tunnel north		Service tunnel		Rail tunnel south
Pumping 21kV feeder 1	Pumping substations (north)	Pumping 21kV feeder 2	Pumping substations (south)	
	Electrical substations (north)	Substations 21KV feeder 1	Electrical substations (south)	Substations 21kV feeder 2
Shafts 21kV feeder 1	–	–	–	Shafts 21kV feeder 2
3.3kV room feeders (north)	Electrical rooms (north)	–	Electrical rooms (south)	3.3kV room feeders (south)

With regard to the four prime 21kV cable circuits, one of the pumping station feeds is located in rail tunnel north and the other is located in the service tunnel. The electrical substation cable feeders are located one in the service tunnel and the other in rail tunnel south. For the underland part between the shafts and the main substations, the two additional cables are run one in each of the rail tunnels.

The equipment located in the pumping substations, electrical substations and electrical rooms has to be very compact because of the restricted space available and must not provide a significant fire risk. This has led to the use of SF_6 switchgear for 21kV, vacuum for 3.3kV and air gear for 400V. The transformers are all dry-type cast resin. The environment in the rooms is relatively dry and so the requirements for special ingress protection are not essential. However, equipment to be located in the tunnels themselves where tunnel washing by hosing down will be used requires IP 65 rating.

Equipment which uses inherently flammable materials is rigorously tested for:

- flame propagation: the materials must not ignite easily nor propagate flame along the tunnel
- smoke density: when burning the material must not produce dense smoke which would obscure visibility and hinder escape
- toxicity: when burning the materials must not produce toxic or corrosive gases.

These types of test apply to all flammable equipment, although the detail of the test is tailored to suit the specific item of equipment under test. All cables to be installed in the tunnels undergo these tests and are classified as low smoke and fume, and have flame retardency characteristics in accordance with IEC 332, the category depending on the voltage rating of the cable concerned.

The cable-supporting system and installation method was designed to meet the following requirements:

- safety under normal and fire conditions
- resistance to corrosion
- minimise installation time in the tunnels
- minimise the earthing requirements
- coordinate with the temporary services.

This has resulted in a system of nonmetallic guttering and trays which have been submitted to the fire testing referred to earlier. This system means that cables are laid directly onto a supporting surface and do not depend on cleating or tying to keep them in place during cable laying.

Because of the large quantities of cables laid and the logistical problems of access from the shaft locations only, an innovative method of cable laying from a slow-moving (0.5km/h) train was developed.

The cable is anchored to the wall at the start point. As the train moves slowly forward the cable is fed out from the drum through a tensioning wagon. This wagon is there to give an alarm if the cable comes under excessive tension. It then passes to the cable feed wagon where it is raised over rollers to the working level where operatives feed it onto the support system.

SHAFTS AND STANDBY GENERATION

The shaft locations are the major load centres for the tunnels, both for essential and nonessential loads.

The distribution system at the shafts is basically on the surface. There are separate 3.3kV and 400V switchboards associated with each of the systems, normal ventilation, supplementary ventilation, firefighting, etc.

In the event that all the supplies from France and the UK are lost simultaneously, standby generators are located at both shaft locations to support the essential services for tunnel evacuation and for the integrity of the tunnel. These loads are essentially lighting, ventilation, pumping, and control and communication power supplies. All other loads have to be shed under these extreme operating conditions.

Use is made of the construction generators already existing at Shakespeare Cliff and Sangatte. The generators are connected via stepup transformers, one per unit at the French side as already existing and one per pair of units on the UK side, to a generator 21kV switchboard.

The 21kV system is a substantial cable network and when energised on no load generates approximately 4.5MVAr onto each set of generators. This self-excitation of the system would cause instability of the generators and it has been found necessary to add a 3MVAr 21kV shunt-connected reactor to each generator busbar to improve the stability of the machines.

In the event of loss of both grid connections the generators are run up, synchronised to each other, loaded with the shunt reactor and then connected to the home board at their location. The network is reconfigured to split both pumping station feeders at pumping station K and then loaded progressively to secure first the lighting, then ventilation, pumping and, last, the control and communication power supplies which have 3h battery-backed autonomy to survive an AC system interruption.

TERMINALS AND EXTERIOR LIGHTING

The terminals are fairly conventional distribution systems with 21kV as the primary voltage and 400V as the utilisation voltage. The UK side uses a ring main concept with duplicate ring main units (RMUs) at strategic substations such as the control centre and single RMUs at other locations. At those locations where only one RMU is provided the 400V switchboard is divided into priority and nonpriority sections. The priority section is secured by a 400V cable interconnection from the adjacent substation. The French side, where the terminal is significantly larger, has used duplicate radial feed systems which can be connected as rings at the terminals.

The exterior lighting consists basically of street lighting, car park lighting, railway sidings lighting and the vehicle loading platform lighting. Most of the lighting is by sodium lamps on conventional lamp posts whereas the platform lighting uses catenary-suspended light fittings.

The lighting design, as well as meeting the functional requirements, has endeavoured to take account of the architectural and aesthetic aspects required by the client and environmental considerations such as minimising glow in the sky, direct light affecting residents and reducing the general forest appearance of lamp posts in daytime, as required by the planning authorities. Certain lamp posts in restricted access areas, e.g. maintenance sidings, have been developed with H-frame luminaire support heads which can be lowered vertically down the support post for maintenance.

The development of the power system for the Channel Tunnel has basically been the application of well-proven and reliable technology to meet the specific and unique requirements of this particular project, with special consideration for load balancing on the UK side and solving the logistical and environmental problems of the tunnels.

CATENARY

An overhead catenary system provides power for the locomotives, a 25kV supply both for the tunnels and the terminals, and also transmits power for BR- and SNCF-operated services. A huge amount of power needs to be supplied to the shuttle trains in the Tunnel; consequently this is one of the most powerful 25kV catenary systems in existence

The catenary was installed by a Spie Batignolles/Balfour Beatty Power joint venture, involving 210 single-track kilometres and 950km of overhead catenary conductors with 14,000 supports.

The large quantity of copper that constitutes the catenary needs to be suspended over the tracks in the Tunnel. This must remain uncovered, and the air space around the conductors that is necessary for safety was difficult to find. The support and other equipment was therefore of very compact design.

The English and French joint-venture subcontractors came from different technical backgrounds, UK electrification and French TGVs, which combined successfully.

For safety reasons, the catenary has to have the facility for sectional switchout every 1.2km along the Tunnel, an unusually high frequency. Although, where possible, already existing, proven components were used for the catenary, some new developments were necessary. Existing isolator designs were modified to withstand the intense pressure from the aerodynamics of the Tunnel. The isolators' movement was transferred from the operating mechanism in the tunnel roof by using control cables usually found in army tanks. The isolators, like all the equipment in the Tunnel, had to be fully waterproof and dustproof.

10

Signalling and train control

PAUL ROBINS

The signalling system within the Channel Tunnel has to cope with the sort of train frequencies normally found on a busy urban passenger railway, with the added complication that a mix of trains with differing maximum speeds and performance must be accommodated. This blend of traffic at fairly high speed and close headways, together with exceptionally stringent safety requirements arising from the length of the Tunnel, has led to many special measures being taken when designing the signalling and train control systems. Some features differ from normal railway practice.

The national railway companies operate high-speed trains between Paris, Brussels and London, running at up to 300km/h outside the Tunnel, but restricted to 160km/h within it. They will also run locomotive-hauled freight trains of various types at speeds from 100 to 140km/h and ultimately 160km/h. The Eurotunnel shuttle trains, 750m long, will operate at a maximum speed of 140km/h, again with an ultimate possibility of 160km/h.

The performance target for the signalling system was a capacity of 20 train paths per hour in each direction, using shuttle traction and braking characteristics to define a path, subject to a maximum speed of 160km/h. This will expand to 24 paths per hour and ultimately to 30, though the last-mentioned target will require a different signalling system.

Under the agreement reached in 1987, these paths are divided equally between the railways and Eurotunnel. However, since the performance of the railways' trains is not the same as that of the Eurotunnel shuttles, each through train may require two or more shuttle paths. Thus in practice the likely initial maximum capacity is 15 trains per hour: 10 shuttles and five through trains.

SIGNALLING SYSTEM

A cab signalling system is used; that is, all indications of adjacent train movements are normally given to the train driver on a display in the driving cab. There is also an automatic train protection (ATP) feature which stops the train if the driver fails to control the train speed as directed by the in-cab display.

The cab signalling system selected was the TVM430. This was developed by SNCF from the TVM300 used first on the high-speed line from Paris to Lyons. TVM430 has been installed on the new line from Paris to the Tunnel (TGV Nord), and its use within the Tunnel avoids the need to equip international trains with an additional set of on-board signalling equipment. It also means that the driver has to be familiar with only three types of signalling (French lineside, TVM430, UK lineside), rather than four.

PHILOSOPHY

TVM430 is implemented in the Tunnel using generally the same philosophy as on TGV Nord, i.e. the codes transmitted to the train determine the maximum permitted speed. The stopping distance is divided into four block sections, in each of which the driver receives instructions to reduce speed. In addition there is an overlap block section, separating the stopping point from the occupied block section, and a warning block, in which the driver is alerted to the need to commence a speed reduction in the next block.

The indication to the driver takes the form of a three-digit, fail-safe display. This gives current safe maximum speed in the case of a 'line clear', his target speed in the case of a speed reduction, or new safe maximum speed in the case of a speed restriction. The distinction is made by the colour of the digits: black on green for 'line clear', black on white for warning of a speed reduction, white on black to instruct execution of a speed restriction, black on red for a stop. In the case where speed reduction must continue in the next block, the display flashes. Should the driver enter the overlap block section, an occupied block, or a section where the transmission of codes from trackside equipment has failed, the display shows red.

The driver is backed by the fail-safe ATP system. If the train exceeds the maximum safe speed at any time by more than 10km/h (5km/h below 80km/h), an emergency brake application is enforced to bring the train to a stand.

The principal difference in philosophy between TGV Nord and the Tunnel concerns the use of permissive blocks. On the TGV Nord line, the blocks are normally permissive, unless there is pointwork within the block section. Thus, after bringing a train to a stand at the block marker denoting the end of the block section, the driver may proceed into the following section under 'proceed on sight' rules without further authorisation. The ATP system restricts speed under these conditions to 30km/h.

All blocks in the Tunnel are absolute, and may not be passed by the driver until authorised to do so. This authorisation is given by a white light mounted on the block marker, which is illuminated by the controller when it is safe to allow the train to proceed on sight, again restricted by the ATP to 35km/h. If the driver attempts to proceed without authorisation an emergency brake application is enforced by the ATP system.

TRACKSIDE EQUIPMENT

TVM430 makes use of the running rails as the signal transmission medium. Audio-frequency jointless track circuits are used both for train detection and data transmission. Four carrier frequencies are available, two of which are used on each running line, alternating from track circuit to track circuit, i.e. $f_1, f_2, f_1 \ldots$ on line 1; $f_3, f_4, f_3 \ldots$ on line 2. The carrier frequency is modulated with one or more from a total of 27 possible low-modulation frequencies. Since any combination of these frequencies may be present at one time, the code transmitted from the track circuit may be thought of as a 27-bit digital word, the absence of any frequency representing a zero in that bit position, the presence representing a one. This word is divided up into fields, each of which represents an element of information required by the on-train equipment. These are: speed code, gradient, block section length, network code, error checking.

The speed code represents in a coded form the current maximum safe speed for the train, the speed it should be doing at the end of the current block (the 'target speed'), and the target speed for the end of the next block. The code transmitted will depend, therefore, on the speed limit in force owing to track geometry constraints, and the state of track circuits ahead. The gradient is the average over the current block, the length of which is specified by the next field.

The network code allows the same track-to-train transmission system to be used on lines with different characteristics. Depending on the network code, the same speed code is interpreted in different ways by the on-board equipment. For example, on the high-speed line the maximum speed is 300km/h, requiring a particular interpretation of the speed codes. In the Tunnel, where the speeds are much lower, a completely different interpretation is required. Thus the high-speed line and Tunnel TVM430 systems transmit different network codes.

The error checking field is a check-sum of the previous fields, allowing the on-train equipment to verify correct reception of the codes.

To provide the line capacities specified, track circuit lengths in the Tunnel are around 450m, and considerably less in the terminals.

Track circuit transmitters and receivers are located with the processors that calculate the codes to be transmitted. These are housed in the Tunnel in equipment rooms at intervals of about 14km, determined by the 7km maximum feeding distance for the track circuits. Processing is carried out by the *monoprocesseur codé* (coded single processor). This device uses a single industry standard microprocessor (Motorola 68000 series) to perform processing to a vital standard (mean time between wrong side outputs $< 10^9$). This is achieved by the coding of the data and checking of the codes generated from these codes during processing. The *monoprocesseur codé* was developed for use in the SACEM signalling system now in service on RER Line A in Paris.

In addition to transmission by track circuit, a nonvital means of transmitting supplementary information is available. This is a cable loop placed in the track where required, which transmits a phase-shift modulated digital signal. This is used in the Tunnel for several functions, including automatic control of locomotive power through neutral sections, and to enforce the emergency stop after unauthorised passing of a block marker.

All temporary and permanent speed limits are imposed through the TVM430 system. Temporary restrictions may be imposed either locally, by switches located at equipment cross-passages every 750m, or remotely through the rail traffic management system at the control centre. In the Tunnel, speed limits of 100, 60 or 30 km/h are available. A 0 km/h speed limit may also be imposed as protection for engineering works.

ON-TRAIN EQUIPMENT

The signal from the track circuit is detected by two pairs of antennae, one from each pair being over each running rail. These are mounted on the locomotive body, about 1m ahead of the leading axle. The signal from each pair of antennae is decoded by a digital signal processor, one per pair. The function of these processors is to demodulate the signal frequencies to produce the 27-bit data word.

The data is fed to a *monoprocesseur codé* forming the display processor, which performs the main decoding functions. If the data received from the two digital signal processors differs, the *monoprocesseur* signals an error, causing the system to shut down. This guards against hardware faults in the signal processors.

The display processor decodes the speed code field and drives the cab display through relays whose position is monitored by the processor. It also passes the speed, together with the gradient and block length information, to a separate processor which carries out the ATP functions. The ATP processor uses this data to calculate a deceleration curve which the train should follow. The real speed is continuously monitored against this curve, and emergency braking is called if the safe speed is exceeded. The real train speed is obtained from a fail-safe tachometer, which obtains three readings from two independently driven axles. It has inputs from the braking and slip/slide protection systems and uses these to produce a speed measurement accurate to 2%.

The ATP processor is not a *monoprocesseur codé*. It guards against hardware error by a different mechanism. Each program cycle is executed twice, using different memory areas and driving separate outputs. A hardware self-test routine is also executed every program cycle. Each output drives a separate relay, whose contacts are monitored by the processor. Contacts of these relays are in the emergency brake circuit, such that, if either drops, the brakes are applied.

All on-board equipment is monitored by a fail-safe monitoring device, which cuts the power supply if a fault is detected. The on-board system is completely duplicated for availability, so an emergency brake application is not made unless both equipments fail. This is because it is highly undesirable for safety reasons to stop trains unnecessarily in the Tunnel; duplication minimises emergency stops caused solely by equipment failure.

INTERLOCKING

Interlocking functions are performed by standard PRCI (poste a relais a commande informatique) equipment as used by SNCF. All vital functions are performed by fail-safe relays, while nonvital control functions are performed by a series of dedicated microprocessors. It may appear strange that, given the availability of BR's proven solid-state interlocking system (SSI), the decision was made to use relay technology. However, TVM430 is designed to work with the PRCI, and vice versa. Use of the PRCI therefore minimised the development risk, an important consideration in a privately funded project with a tight construction schedule. Furthermore, in a tunnel environment, where it is not desirable to distribute active equipment along the trackside, the cabling economies normally offered by SSI are not obtainable.

CONTROL CENTRE EQUIPMENT

There are two control centres, one at Folkestone, the other at Calais. Normally, that at Folkestone is in control, and is thus the more fully equipped of the two. The Calais control centre is capable of taking over all the functions of that at Folkestone in the event of failure or other incident.

The control centre system consists of two subsystems: the rail traffic management (RTM) and the engineering management system (EMS). The function of the first-mentioned is to act as the operator interface to the signalling, providing the usual route-setting and train description functions, along with some timetabling and automatic route-setting facilities. The last-mentioned monitors and controls all the nonsignalling functions: traction and other power supplies, ventilation, pumping, cooling, fire detection and suppression, and the condition of all control and communication equipment.

In view of the requirement for very high availability, the UK control room is also equipped with a VDU and keyboard giving direct access to the signalling system, providing basic route-setting functions. The control room mimic diagram is also driven directly by the signalling system. Thus, in the event of a failure of the RTM system, sufficient controls and indications are available, at the very least, to ensure safe and rapid evacuation of the Tunnel.

RAIL TRAFFIC MANAGEMENT

The RTM system provides each operator with colour VDU displays of three types: track diagram, text and train graph. These are arranged, together with two VDUs for EMS functions, at five workstations in the UK control centre and three in France. A supervision position is also provided, at which summary RTM information is available. Operator input is by tracker-ball and keyboard. The track diagram provides, with a selectable level of detail, indication of train positions and descriptions, routes set, signal states, temporary speed limits and lines blocked for engineers' possessions. The text VDU is used for input of instructions and presentation of alarms and other messages. The train graph VDU provides the operator with a real-time graph showing the recent and projected progress of all trains in the Tunnel and terminals. The system detects potential conflicts automatically and alerts the operator.

In normal circumstances, trains are routed automatically by timetable. Three levels of timetable are produced: the long-term schedule, the tactical schedule and the adjusted schedule. The long-term schedule is prepared off-line, weeks or months in advance, and represents the intended basic service for any given day. This is modified daily to produce the tactical schedule, covering several hours of traffic, which gives the real service including, for example, trains for which 'ghost' paths have been included in the long-term schedule.

It is this schedule that forms the basis of the automatic route-setting system. There are, of course, disruptions and unexpected events which mean that the tactical schedule is not adhered to. The operator uses the train graph to identify these events and their effects, and prepare strategies to minimise their impact. These strategies may be tested on the train graph before putting them into effect by modifying the tactical schedule to produce the adjusted schedule, which is used to determine the routes actually set.

This philosophy is intended to encourage operators to think and plan ahead, rather than behave reactively to events on the ground. Normally, operators should not have to resort to issuing direct route-setting commands, though these are, of course, available.

INTERFACE WITH NATIONAL RAILWAYS

The control centre manages all services through the Tunnel, and allocates paths to both the shuttle fleet and through trains. All freight services originate at the Fréthun or Dollands Moor yards, from which they can be called forward when Eurotunnel is ready to accept them. Overnight passenger trains also stop in these yards, where locomotives are attached or detached. For Eurostar trains, the control boundary more or less coincides with the limits of Eurotunnel''s Concession. On the British side, this is where the four-track Continental main line crosses the M20 motorway. Here, control of the trains trains is handed over to or received from Railtrack's new signalbox at Ashford, which controls Dollands Moor yard and most of the principal line to London via Tonbridge.

In France, a new SNCF signalbox at Fréthun regulates all train movements to and from the Eurotunnel Concession tracks, the boundary being even closer to the portal than it is at Folkestone.

11

Terminal and tunnel trackwork

MICHAEL BAXTER AND PETER DAVIES

A decade after opening, the annual gross tonnage carried in each direction through the Tunnel, according to design assumptions, may exceed 100 million tonnes per annum. The only known railways with this tonnage elsewhere in the world are heavy-haul lines in the USA, such as Powder River basin in Wyoming and the Union Pacific Railroad of Nebraska. However, the ultimate traffic level forecast is 240 million tonnes per annum in each direction, which is probably the highest load imposed on any track since railways began!

Table 1 gives a breakdown of the design traffic levels by train type in 2003 and the ultimate capacity. Shuttles account for 55% of trains and 73% of gross tonnage in 2003; this increases to 71 and 85%, respectively, as ultimate capacity is reached. The ultimate capacity of 130,875 trains a year in each direction assumes the signalling system will permit a 2 min headway.

Table 1. One-way traffic design assumptions made to determine track load to be carried in 2003 and when ultimate capacity at 2 min headway is reached

Type of train	Speed (km/h)	Axleload (t) Locos	Axleload (t) Wagons	Capacity in 2003 Trains	Capacity in 2003 mt	Ultimate capacity Trains	Ultimate capacity mt
Tourist shuttles	140	22.5	22	18,572 }	74.8	59,216 }	203.1
Freight shuttles	140	22.5	22	17,595 }		34,309 }	
Eurostars	200	17	17 }	20,000	15.0	25,200	18.9
Hauled passenger	130	22.5	12 }				
Fast freight	130	22.5	22.5	3,600	3.8	4,860	5.2
Slow freight	100	22.5	22.5	5,400	9.5	7,290	12.8
Totals	-	-	-	65,167	103.1	130,875	240.0

Figure 1 overleaf shows the basic configuration of the terminal and tunnel trackwork.

TERMINALS

The trackwork in the British and French terminals (Tables 2 and 3) is generally conventional ballasted track with twin block sleepers at 580–600mm centres

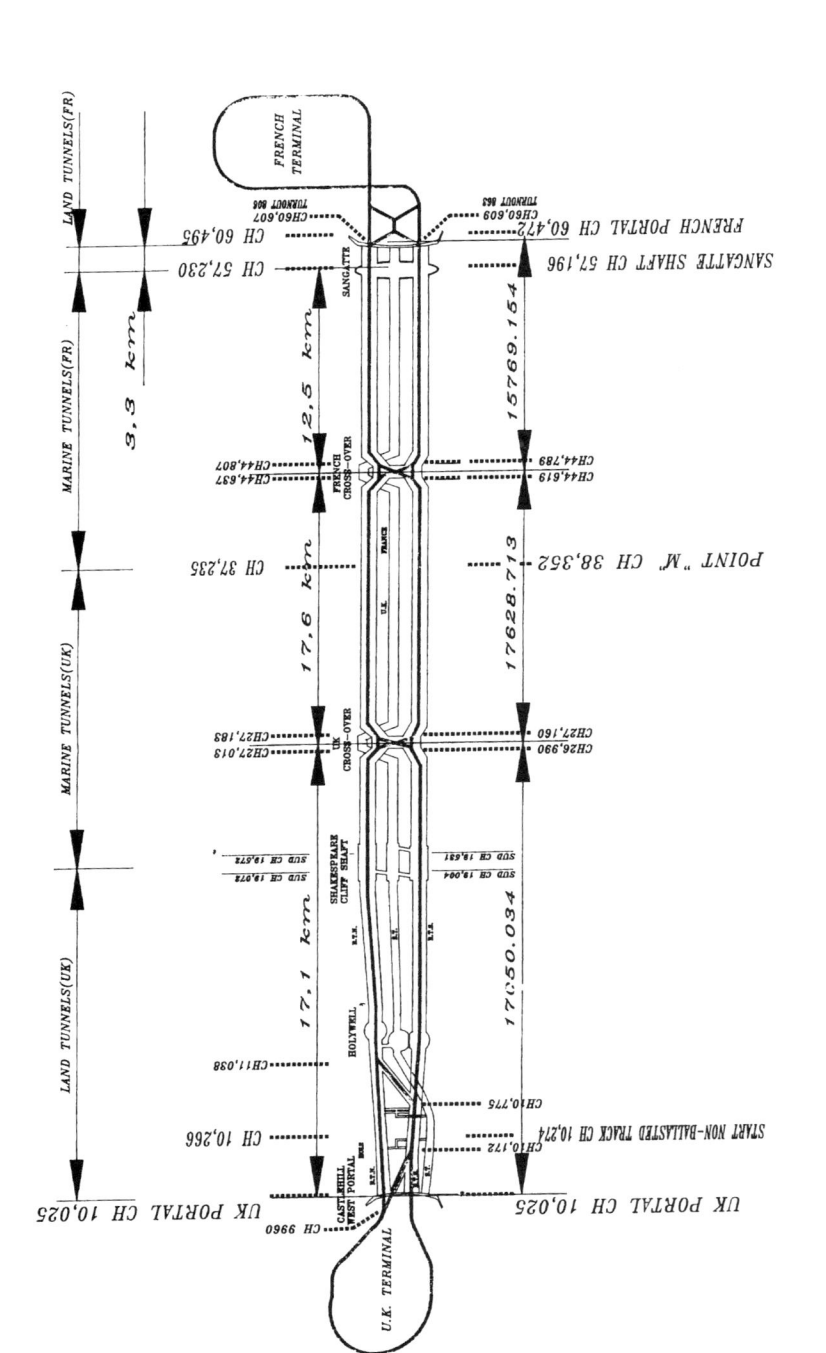

Figure 1. Schematic layout of Channel Tunnel terminal and tunnel trackwork, showing locations of crossovers, portals and main interfaces

and standard SNCF-type Nabla rail fastenings, laid on 350mm ballast stone. This was placed on a 200mm layer of sub-ballast; sand blankets were used where gault clay outcrops occurred. The platform tracks are placed on ballast within a concrete trough along the full length. Concrete upstands are used for propping the shuttle wagons during loading and unloading. A smaller twin block sleeper (U31) was used to fit within the trough. Turnouts were fabricated on azobe hardwood timbers from West Africa. In the workshop areas, embedded rails are used through the concrete approach aprons to provide at-grade crossings of the tracks for pedestrians and vehicles.

The shuttle trains traverse a figure of eight through the terminals to reduce uneven wear on the rolling-stock wheels; they travel clockwise through the Folkestone Terminal system and anticlockwise through the Calais Terminal.

Table 2. Main terminal trackwork parameters

Item	UK terminal	French terminal
Main line UIC60 rail, km	20	30
Main line BR connection 113A rail, km	7	
Secondary lines U50 rail, km	11	20.5
Turnouts UIC60	45	44
U50	22	42
113A	11	

Table 3. Terminal trackwork types

Item	Main line	Secondary lines	BR connection
Concrete sleeper	Twin block	Twin block	Monoblock
Type	U41	U31	EF 28S
Platforms	U31		
Resilient pads, mm	9	9	5
Fastenings	Nabla	Nabla	Pandrol
Timber sleepers	2600x260x180 adzed flat	2500x260x150 adzed 1/20	
Resilient pad, mm	4.5	4.5	
Baseplate	1/20 slope	Flat	
Fastening	Nabla spring with T-bolt	Nabla spring with coach screw	
Emergency siding (UK)		Base plate (U50) on concrete	

The Nabla fastening for the concrete twin block sleepers comprises a single-leaf spring seated on an insulator placed over a cast-in holding-down stud which is set in the concrete; the spring is tensioned against the rail foot by a torqued nut, placed in the stud. For timber sleepers, the stud and nut arrangement is replaced by a sleeper screw. The timber sleepers on the main line were adzed with a level surface but with the rail seat for the cast baseplates at 1:20 inclination; the secondary track sleepers were adzed with a surface slope of 1:20 to the centre for flat baseplates.

Figure 2. Sleepers being laid in the M20 rail bridge between the Folkestone Terminal and Dollands Moor holding siding for freight. (Source: QA Photos)

Platforms are 840m long, and there is provision for additional island platforms at each terminal to cope with future traffic demands. The terminal layouts have been designed to allow for rerouteing if a turnout is inoperable.

All curves within the terminal areas with a radius of less than 1000m are laid on azobe timbers between 500 and 550mm spacing so that check rails can be retrofitted should this be required. Portec P100-type lubricators, with vibratory sensors for activation, are located on curves below 750m radius.

For the general design and layout of the terminals, see Chapter 19.

Folkestone Terminal

In the Folkestone Terminal, there are approximately 20km of main-line track and 45 turnouts serving eight platform tracks and one emergency siding, which is designed for firefighting or bomb disposal on trains. This siding is constructed on slab track, incorporating baseplates bolted to a concrete base to avoid the risk of flying ballast stones should a bomb be detonated on board a train. There is provision in the track fan layout for eight more tracks to serve the future platforms when the capacity of the present layout is reached.

The shuttle loop consists of two tracks which diverge to the left from the London-bound Continental main line (CML) from just inside the Channel Tunnel portal. They continue along the southern perimeter of the terminal and pass below the CML in a cut-and-cover tunnel where a third track provides access to the servicing area. The tracks curve to the right around terminal buildings through an angle of over 200° on 300–420m radius curves, to emerge at the west end fan leading into the platform tracks. After converging again to the east of the platforms the loop joins the Paris-bound main line at the portal. Alongside the terminal, the CML is provided with a third track between the up and down lines, which can be accessed from either, for reversing trains before the portal. The train servicing area between the CML and the shuttle tracks to the south has 11km of plain line with U50 rails and 22 turnouts.

Tracklaying, undertaken by Balfour Beatty Railway Construction under a subcontract, commenced in December 1989 and was completed by the autumn of 1992. Figure 2 shows sleepers being laid at the Dollands Moor siding.

Calais Terminal

The length of main-line track in the Calais Terminal is approximately 30km, with 44 turnouts. The loop extends for over 1km on its approach to the platforms; beyond, the track passes through a grade-separated structure and then connects into the TGV main line from Lille in the Beussingue cutting which forms the approach to the French tunnel portal. Between the main lines and before the portal a third (central) track provides both crossover and train-reversing facilities. The secondary lines to servicing areas extend over 20.5km of plain line and 42 turnouts. The reason for the difference in overall length of tracks between the Folkestone and Calais Terminals is that more space was available; thus the main train servicing facility is at Calais. As at Folkestone, an emergency siding is also provided but, as it is contained between embankments and is remote from other facilities, normal trackwork on ballast was used.

Tracklaying, undertaken by Spie-Batignolles under a subcontract, commenced in December 1990 and was completed by summer 1992.

TUNNELS

Ballasted tracks extend from the Folkestone Terminal approximately 250m beyond the portal through the tunnels below Castle Hill, and include a facing crossover. Due to the very long-term possibility of slight movement of these tunnels within the gault clay, the tracks may be required to be realigned sometime in the future. The tracks are therefore laid on ballast stone placed on the tunnel invert, and the walkway clearances have been increased accordingly. Throughout the rest of the tunnels a nonballasted trackform was used.

The specification for the tunnel track required that concrete supports should last 50 years, and the spring fastenings 25 years. Elastic components were required to last two rail lives, and rail pads and insulators one rail life.

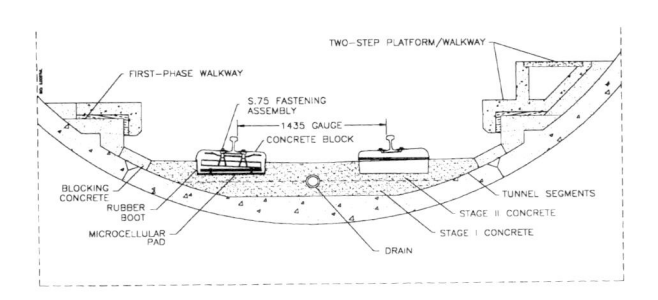

Figure 3. Cross-section of base of rail tunnel

Given the limited access for maintenance which the expected high traffic levels impose, and the need for geometric stability, ballasted track in the tunnels was ruled out at the inception of the project. Other important requirements in selecting track for use in the rail tunnels were:

- correct vertical resilience to be maintained throughout the life of components
- low aerodynamic resistance to the air flow between the track and the underside of trains
- good electrical resistance against leakage of signalling currents carried in the rails
- a rail fastening that permits minor adjustment of track alignment, and allows for future use of rail of heavier section than UIC60
- strict tolerance on gauge widening under combined vertical and lateral loads
- economical to maintain and renew components.

Five different track systems were considered: IPA from Italy, DB/Rheda from Germany, VSB-Stedef from France, PACT from Britain, and a design put forward by Sonneville International Corporation based in the USA. The VSB-Stedef, PACT and Sonneville designs were subsequently selected for rigorous qualification testing.

Based on the results of these tests, and other design parameters, the Sonneville track system was chosen. This incorporates a twin-block rail support consisting of independent concrete blocks under each rail. These blocks rest on resilient microcellular pads inside rubber boots cast into a concrete base, which is laid in two stages on the tunnel invert (see Figure 3).

The principal tests to demonstrate that proposed track systems met the specification were carried out by Dr Ing. J. Eisenmann at Munich Technical University. He considered the primary criteria for good performance to be: an

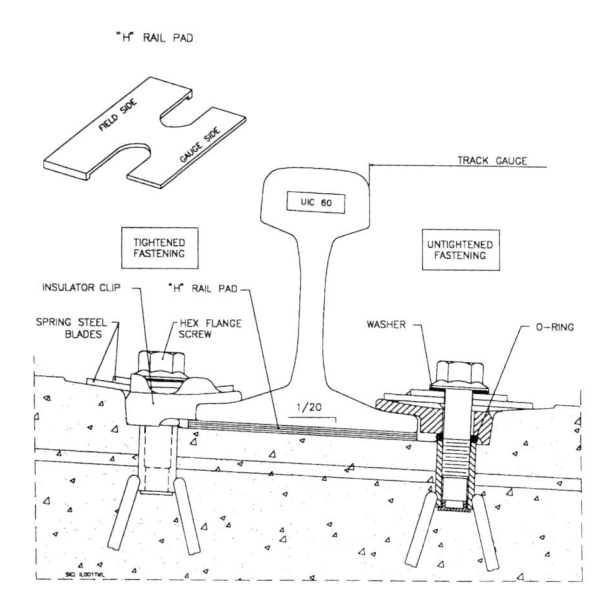

Figure 4. Sonneville system for fastening rails in the Tunnel

acceptable static and dynamic elasticity of the whole track system; resistance to fatigue; and no longitudinal rail creep.

Low static and dynamic spring coefficients are desirable to distribute the load onto several rail supports, and to attenuate dynamic forces; 20–41kN/mm for one rail seat was the range specified for the static loading conditions. The Sonneville system showed an increase in static stiffness of only 5% after 3 million load cycles.

The increase in stiffness which occurs when the same load is applied dynamically at 20Hz is also important. The Sonneville track yielded a particularly low ratio of static-to-dynamic stiffness of 1.2, which was considered to be very good result. After successful completion of the fatigue test the spring coefficient was 32kN/mm, well within the specification.

Track contracts

Sonneville's consultancy with TML covered design and development of the track system and the method of installation, together with advice on quality control throughout manufacture, assembly and installation.

Contracts for supply of the track components were placed by TML direct with various manufacturers. A contract for assembly and installation of track in the rail tunnels was awarded to a consortium known as the Channel Tunnel Trackwork Group (CTTG) comprising Montcocol (France); Tarmac (UK); Borie SAE (France); Travaux du Sud Ouest (TSO, France); and Heitkamp (Germany). Contractual and legal agent for CTTG was Montcocol, which also acted as technical agents.

Montcocol's contract included placing concrete for the stage 1 track base and walkways on the French side, after removal of the construction track, and placing the stage 2 infill concrete throughout the tunnels. Montcocol also installed the precast concrete sections forming the walkways (see Figure 3) which had to be located within ±15mm relative to the rails.

The Sonneville system

UIC60 rails of 900A grade rest on pads 6.2mm thick made of a microcellular EVA material. These pads are not rectangular but H-shaped, as shown in the top left corner of Figure 4. The shorter and narrower side of the pad is placed under the gauge side of the rail, while the larger limb with its locating lips supports the field side, which is subjected to higher loading; shaping the pad in this way improves its performance over a longer service life.

The rail seat is inclined at 1:20 within a recess in the top of the concrete block. The rail is located laterally by insulating nylon clips which bear against the sides of the recess and the rail foot. In-service adjustments to gauge can be effected by replacing these nylon clips with a matched pair of unequal width.

Electrical insulation between rail and block is provided by the nylon clips and microcellular EVA rail pads. The boots surrounding the blocks provide a second line of defence against electrical leakage. This is important in view of the need to preserve the integrity of track circuits carrying cab signalling commands in a humid, saline atmosphere.

The S75 fastening comprises a nylon clip, a twin-leaf spring, a washer, an O-ring, and a screw torqued into a steel insert embedded in the concrete block (see Figure 4). It develops an average toe load of 17kN, or 34kN per rail seat.

It is obviously important that the thread should be protected from the corrosive atmosphere in the tunnels. This was achieved by applying anticorrosion treatment to steel components, packing the inserts with grease on assembly, and sealing them with the rubber O-ring round the screw that is nipped between the clip and the block on tightening. Salt spray tests carried out at UMIST in Manchester revealed no corrosion after 1000h.

The combination of limited gauge widening under load and the high level of vertical resilience specified required a second pad under the block. This is a simple rectangle, 12mm thick, again made of microcellular EVA material. It also provides vibration insulation to limit the amount of noise radiated within the Tunnel.

The lower part of each block, together with its supporting pad, is encased within a rubber boot that extends above the surface of the stage 2 concrete. The vertical surfaces of the boot are ribbed on the inside to provide lateral and longitudinal elasticity, and to avoid wear in the boot that would otherwise result from vertical movement of the blocks. Most of the 335,000 blocks were manufactured in the precasting yard at Sangatte used to make tunnel lining segments on the French side. Close tolerances were specified: a specially designed casting machine manufactured by Casagrande of Italy, and a batching plant and ancillary equipment were supplied to TML by Sonneville. Approximately 100,000 blocks were subcontracted by TML to Costain and manufactured in Tallington, Stamford.

Fastenings were ordered from various suppliers, while the pads and boots were from Saar Gummiwerke and Phoenix of Germany.

Crossovers

Crossovers are required within the tunnels to provide flexibility of train operations during track or tunnel equipment maintenance, or in an emergency. In addition to the facing crossover inside the UK portal, laid on ballast, there is a trailing crossover on Sonneville track before Holywell, approximately 750m from the portal, constructed within the cut-and-cover tunnels. The turnouts are designated UE15 comprising 1:15 angle common crossings. All turnouts within the Sonneville track sections were laid on conventional azobe timbers, each on resilient pads surrounded by individually moulded rubber boots, of the same type used for the Sonneville blocks.

There are two undersea scissors crossovers, designated UK crossover (at 17.1km from the UK portal) and French crossover (at 15.8km from the French portal). They extend over 165m and comprise four UE15 turnouts with 1:15 common crossings and a diamond crossing. To avoid conflict with the trackwork, the two 95-tonne steel doors, which separate the north and south tunnels through the undersea caverns, are designed to cantilever over the diamond crossing in their closed position. To enable the support rollers to travel as close as possible to the back of the inside of the V of the diamond, the diamond crossing and the approach tracks are laid on Edilon Corkelast elastomer embedded within the concrete trackbase. The track resilience and deflections under dynamic loading through the crossovers were designed to be similar to that provided by the Sonneville track in the main lines. The use of the elastomer avoided the need for rail support components such as baseplates and spring clips, which would have prevented the large support beam for the steel doors from being installed against the legs behind the diamond V. It also provides a flush surface between the running rails and the concrete base, which is ideal for cleaning and complies with safety requirements.

Turnouts

Turnouts are designed according to the operating speed expected and basically comprise French standard components. The 1:20 inclination of the rails is continued through the stock and switch rails, thus eliminating the need for twist rails. The crossings are of cast manganese with normal rail-steel legs welded on by a special process in the factory, thus allowing the crossing to be welded into position by the use of *in situ* aluminothermic welds. The switchblades are of asymmetrical section machined in the normal manner. The steelwork of the turnout was mounted on azobe creosoted hardwood timbers adzed to take the various baseplate fittings. Due to the divergence of the tracks this means that every timber in the turnout differs and thus concrete bearers were considered to be uneconomical. Fabricated baseplates were used rather than heavy castings and, to ensure stability, gaugeplates together with rail anchors were installed.

As the UK terminal was originally designed to UK standards the turnouts normally have UK geometry with French components. Where possible, direct replacement with French-type turnouts was carried out.

Due to the spread of turnouts in the terminals, gas point heaters have been provided for terminal trackwork. They are controlled by thermostats which automatically turn the gas on and off depending on the temperature of the rails. The BR section of the CML is equipped with electrical point heating. Expansion switches were also provided in accordance with SNCF practice to a Belgian design in the terminals, and to BR practice and design in the CML connection. A summary of the turnout types used in the terminals and tunnels is given in Table 4.

Trackwork installation

A trackwork installation subcontract, valued at over £150m, was awarded to the Channel Tunnel Trackwork Group (CTTG; see earlier) in October 1990. The subcontract was effectively split into separate UK and French contracts owing to the ventilation and security seal at 'point M' (the junction point between the UK and French tunnels) and the greater scope of work in the French tunnels. (Although the French contractor had less track to lay than its UK counterpart, its task also included the clean-out.) Tracklaying commenced at the UK portal in July 1991 and at the French portal in October 1991. The length of track in the tunnels was 54.5km (UK) and 46km (French).

In the UK tunnels, TML carried out the tunnel invert cleaning (including removal of the 900mm narrow-gauge construction railway, which added to the logistical problems), laying the drainage system and placing the stage 1 concrete ahead of the tracklaying. The tracklaying subcontractor was

Table 4. Turnout types

Turnout type	Turnout radius (m)	Turnout speed (km/h) Cant deficiency 60–70mm	Cant deficiency 80mm
Tg 0.0336 (1/29.74)	3000		110
Tg 0.0312 (1/26.85)	2500		105
Tg 0.0476 (1/21)	1540	88	
SUG 21	1265	80	
UF 18.5	982	70	
Tg 0.0653 (1/15.3)	820	65	
UE 15	645	57	
Tg 0.085 (1/11.8)	599	50	
UD 10.75	332	42	
Tg 0.11	249	35	
UIC60 & 50			
British Rail			
GV 28	1649		100
FV 24	980		80
SGN 18.5	1906 ⎫		
FV 18.5	794 ⎭	60 on curve	
DV 15	330	40	

responsible for locating and placing all cross-track ducts prior to pouring the phase 1 concrete as well as the construction of the evacuation and inspection walkways. In the French tunnels, the contract included installation of all the trackbase works, and the walkways.

The large scale of the operations, the difficult logistics of transporting all materials and operators through the portals, and the tight programme, demanded a large degree of automation. This resulted in probably the greatest concentration of mechanical plant ever used on a tunnel trackwork installation contract.

The design of the tracklaying equipment was based on a nominal rail length of 180m. On the UK side, five 36m lengths of UIC60 rails were flash butt welded at BSC Workington and delivered on modified BR rail trains. After unloading by portal cranes, the rails were stockpiled at CTTG's works area in the Folkestone Terminal adjacent to their assembly workshop. Blocks were fitted on the moving frame located outside the 'boot factory' (see later).

UK tunnels

From the UK portal, tracklaying proceeded down land tunnel north as far as the foot of the Shakespeare Cliff access shaft. As engineering works were still in progress at this location, tracklaying equipment was transferred to land tunnel south portal in January 1992 from where tracklaying proceeded until

Figure 5. Four 180m rail strings for tunnel track are carried through the UK undersea crossover on the rail train. (Source: QA Photos)

both land tunnels were completed. Tracklaying then transferred back to marine rail tunnel north and proceeded to the undersea crossover (Figure 5) where the turnouts had already been fabricated. After installation of the rail lengths between the turnouts, tracklaying proceeded on the final 10km to 'point M'.

Following erection of a temporary lifting bridge across marine rail tunnel south to provide continuing access for the remaining construction railway in the service tunnel from the Shakespeare Cliff construction platform via adit 2, tracklaying proceeded until reaching 'point M' at the end of July 1992. Stage 2 concreting was completed in October. The final connection between the UK and French tracks was made in January 1993, which coincided with the completion of tracklaying in the French tunnels.

As tracklaying progressed, access to the tunnels became more congested owing to the installation of services using specialised rail-mounted plant behind the tracklaying equipment. The tracklaying subcontractor required access for the following trains and mechanical plant:

- rail train
- steel formwork (walkway) handling train
- walkway concreting train

- Framafer RND machine (for track alignment)
- stage 2 track concreting train
- walkway drilling train (installing dowels for precast units)
- precast walkway unit installation train
- vacuum cleaning train.

Due to the different types of operation, tracklaying (placing of rail strings on the stage 1 concrete) and stage 2 trackbase concreting proceeded at varying rates. Generally, four pairs of rail strings (equivalent to 720m of track) were laid in a 12h shift; under exceptional circumstances eight pairs were laid in a 24h period by reloading the rail train and returning it to the railhead during the night shift.

French tunnels

Tracklaying in the French tunnels differed in several respects from that on the UK side. In addition to tracklaying, the subcontractor was also responsible for:

- removal and disposal of the construction railway track and its steel support structure (this was achieved by means of a specially developed machine called 'Diplodocus')
- tunnel invert clean-out using high-pressure water jets
- installation of drainage pipes and manholes
- placing of stage 1 tunnel invert concrete
- drilling of dowel bars into the tunnel segments for the walkway concrete.

The subcontractor developed special mechanical plant equipped with rubber tyres to run along the tunnel invert. Wheels were set normal to the invert (splayed at approximately 30° to the vertical plane through the centre of the machine). Steering was achieved by means of guide wires along the horizontal tunnel axis. At the front end, plastic drainage pipes were fed through a guide orifice which lowered them into the invert; at the rear, stage 1 concrete was placed over the pipes and across the tunnel to form the base. Simultaneously, the *in situ* concrete for the walkways was slip-formed at the side of the machine. Concrete was mixed *in situ* in the concreting train, and fed by a rubber-tyred shuttle (also fitted with splayed wheels) using the same system developed in the UK tunnels.

Lifting and final lining of the track, and placing of the stage 2 concrete, were carried out using the plant previously used in the UK tunnels. Tracklaying commenced in October 1991 from the rail tunnel north portal and was completed in the south tunnel in January 1993.

Component assembly

In the workshop, known as the 'boot factory', the various components of the Sonneville track system were brought together using a mixture of automatic and semi-automatic plant. Separate conveyors carrying concrete rail support blocks, resilient pads and rubber boots converged. Mechanical arms opened out the boot, allowing the resilient pad and concrete blocks to be lifted in by suction pads.

The block-pad-boot combination was transported along another conveyor where the H-profile EVA rail pad was placed on top of the block. A horizontal adhesive plastic tape was automatically wrapped around the top of the boot to prevent ingress of concrete during installation. Two fibre-reinforced tapes were placed vertically around the extremities of the block, and one at the centre around the H-pad. Initially this second operation was carried out manually, but was later automated.

The final operation in the workshop was to place a timber strip below the boot to protect the underside during handling and storage. This was secured to the block and boot by two vertical bands installed by machine. Assembly took about 2 min.

At the rear of the workshop, two narrow-gauge tracks set on a concrete base between the workshop and the rail stockpile carried a transporter frame, approximately 200m long, mounted on flanged rollers. The frames were driven along the tracks by rack-and-pinion. Controlled movement of the frames ensured that the blocks were loaded on it at 600mm centres as they emerged from the workshop.

Starting at one end of the frame, blocks were spaced along it until sufficient were in place to support a 180m-long rail. Pre-assembled S75 rail fastening components were loaded on both sides of the rail seat on each block. When the full complement of blocks had been placed the frame was directly opposite the rail stockpile. A rail was then lifted by the overhead gantry cranes, and placed directly onto the rail pads on the blocks. The fastenings were screwed manually into steel inserts in the blocks, with the nylon insulator bearing on the rail foot.

The frame then reversed past the workshop with steel rollers pushing on the rail web to close up any field-side gaps on the fastenings. The bolts were tightened using air-driven torque spanners. After checking, the frame was again reversed. The resulting 'rail string' was lifted from the frame and carried over the rail stockpile onto the string stockpile by the portal cranes.

Tracklaying

The rail string train on the British side, approximately 400m long, consisted of 20 modified flat-wagons. These carried secondary support frames, mounted on small rollers, on which the rail strings were laid. These frames allowed small relative movements of the wagon to take place when traversing curves and turnouts without inducing stresses in the rail strings. Eight strings were laid on the wagons in two groups of four, one behind the other on the train, i.e. sufficient to lay 720m of track.

At the railhead, the tracklaying sequence started with the ten leading wagons standing on a 200m length of portable track spanned by nine portable gantry cranes, each with a pair of spreader beams. The two spreader beams on each crane were lowered onto two of the four rail strings on the wagons; gripper arms closed below the railheads, giving 27 support points to each rail string. The two strings were then lifted off the train, and held above it.

The train withdrew, and fishplates securing the portable track section to the end of the permanent track were removed. The portable track, formed of secondhand rail and timber sleepers, was towed forward for 180m along the concrete invert by a Unimog road/rail tractor. Sprung rollers lifted the track off the concrete during the towing operation; the springs retracted when the rail train moved onto the track, causing the sleepers to rest firmly on the concrete base.

Once the portable track had been pulled clear, cranes lowered the two strings of rail onto the stage 1 concrete trackbed. The cranes were programmed to place the rail strings within a few millimetres of their final alignment. The ends of the rail strings were cut to ensure the required gap width for subsequent *in situ* aluminothermic welding. Temporary fishplates and gauge (tie) bars were installed to enable the rail train to be propelled onto the new length of track, which was fishplated to the portable track at the other end.

The next task was to move the gantry cranes forward by 180m so that they could lift and place the next pair of rail strings. The locomotive again positioned the leading ten wagons beneath the cranes. Arms mounted on the gantries engaged brackets on the sides of the wagons, and the hydraulic feet were withdrawn so that the weight was transferred to the wagons. The train then moved forward onto the portable track until the cranes were positioned correctly for the next lift. The next tracklaying cycle could then begin. Rail strings on the second group of wagons were lifted two at a time by the gantries, placed temporarily on the leading wagons, and then laid in position in the normal way after the gantries had been carried forward. Figure 6 shows the gantry in operation.

The whole operation of placing a pair of strings and then moving the gantries forward took about 2h, enabling up to four pairs of rail strings to be laid per day, equivalent to 720m of track.

Figure 6. Tracklaying in the UK crossover: gantry arms move across and down above completed track

In situ concrete for walkways

The next major operation was placing *in situ* base concrete for the walkways on both sides of the welded tracks. Short lengths of steel reinforcement bars were drilled into the haunches on the concrete lining segments; these bond the walkway concrete to the tunnel lining. Later, shutters were installed outside the trackform, resting on the tunnel invert. Cross-track ducts were installed under the walkways and between the Sonneville blocks, prior to concreting.

Logistically, placing the walkway concrete presented the greatest challenge to the subcontractor. The need to pour over 370m^3 of walkway base in a day up to 27km from the portal required a train 580m long weighing over 2000 tonnes when fully laden. Hauled by three diesel locomotives, this was the heaviest train required during the construction works.

The 15 hopper wagons were linked by conveyor belts which fed aggregate into a large mixer where cement and water were added under computer control. The train discharged into the areas behind the shutters, providing a base for the precast units. Evacuation walkways, incorporating cable troughs, are located on the 'inside' of the tunnels, providing access to the cross-passages and the service tunnel in an emergency scenario. Maintenance walkways are located on the 'outside' of the tunnels.

Final track alignment

To comply with the tight absolute and relative tolerances of the final track alignment a 'top down' system was used. To lift the rails and support blocks up to final level, a Framafer RND 92 machine running on the track laid on the phase 1 concrete, raised the rails behind it using grippers and hydraulic rams bearing on the concrete base. Prior to the lifting operation, the temporary gauge bars had been replaced by heavy section bars, at 1.8m centres, which incorporated screw jacks and placed an inclination of 1:20 on the rails.

The position of the Framafer machine was set by two infrared beams projected from a theodolite positioned on the trackbase using survey stations. The machine was linked to the theodolite through a control cable and a computer.

When the location of the machine along the track had been computed, the designed values of x, y and z coordinates of the track in space were calculated. By means of prisms reflecting the light beams back to the source, the actual position of the machine and track could be obtained. When the track had been lifted to its final line and level, the operator used a pneumatic torque spanner to turn the screw jacks to maintain the correct level. Props bearing on the walkway concrete were manually adjusted to place the track on its required alignment. Next, the timber strips were removed from the boots. Cycle time for this operation was approximately 2 min. Then, a manual check using conventional levelling equipment was carried out. Finally, a four-axle light-weight recording frame (CM10) was pushed along the lined and levelled track. This produced a hard copy continuous trace of cant, versine, level and gauge. Any final adjustments were carried out prior to the placing of the stage 2 trackbase concrete.

Stage 2 concreting

Placing the stage 2 trackbase required 210m^3 of concrete a day to be provided using a 460m-long train. This was propelled to the end of the previously concreted and cured track. To distribute the mixed concrete over the lifted and aligned track, a huge 12-wheel rubber-tyred shuttle ran on the *in situ* concrete for the walkways between the concrete train and the placing unit, which was also mounted on rubber tyres.

Concrete was transported in a 4m^3 container inside the shuttle. At the discharge chute on the train, the container was lowered to receive the mix.

Figure 7. Completed phase 2 concreting in UK crossover, with Montcocol concrete train

Prior to placing, the container was raised by hydraulic rams to discharge into the receiving hopper. Both container and hopper were equipped with agitating screws.

From the hopper, the concrete fell under gravity into a main discharge chute located between the rails and minor chutes alongside the Sonneville blocks. An operator controlled the flows to ensure the correct distribution of concrete around and below the blocks. A fourth chute topped up with concrete at the rear. Poker vibrators were located near the base of the chutes. Profiling, levelling and smoothing of the concrete surface was carried out manually behind the train. Finally, a curing compound was applied. After setting, which took approximately 18h, the jacks were unscrewed and removed from the pockets in the concrete along with the gauge bars. The pockets were later filled with grout.

As the concrete trackbase is not reinforced, its lateral stability depends on the anchoring effect of the *in situ* walkways. To transmit horizontal forces from the track, 'blocking concrete' was placed in the recesses between the trackbase and the *in situ* walkway by means of the chutes at the rear of the concrete train. This completed the trackbase profile. Precast walkway units were placed on the *in situ* concrete on both sides of the track. The evacuation walkway units were designed to a profile incorporating a continuous trough for communication and signalling cables. Precast covers were bolted on top of the recess to form the walking surface. The inspection walkway units are a simple step design. Figure 7 shows completed phase 2 concreting.

Because of the complex section, which follows the profile of the shuttle train structure gauge, and the tight installation tolerances required for the walkway surfaces, precast units were specified. Anchor bolts hold these units in place prior to grouting behind and under them. These walkway units are also designed to contain a derailed shuttle train within the confines of the trackbase.

# 12

Locomotives

ROGER FORD

Two fleets of specially commissioned, powerful six-axle locomotives were ordered to haul trains through the Channel Tunnel. These are, first, Eurotunnel's own locomotives for its passenger-vehicle and freight shuttles; and, second, the Class 92 locomotives that haul the European Night Services and freight services. In addition, the national railways commissioned a fleet of Eurostar trains; these substantially modified TGVs operate international services through the Tunnel between London (and beyond) and Paris/Brussels.

For the shuttle and Class 92 locomotives, Brush Traction of Loughborough teamed up with ASEA Brown Boveri (ABB) in Zurich to secure the contracts. Broadly, Brush assumed responsibility for the mechanical portion, while ABB designed and supplied the electrical equipment. All these locomotives were assembled by Brush at Loughborough.

First to be completed and enter service were the 38 locomotives for Le Shuttle. Rated at 5.6MW, these always work in pairs, one at each end of a shuttle train, and the superstructure is not symmetrical. They were followed by 46 locomotives with the conventional arrangement of a full-width cab at each end. Designated Class 92, these are to haul freight and overnight passenger trains through the Tunnel but they must also work on both 25kV AC overhead and 750V DC third-rail lines in the UK. They are rated at 5MW on AC and 4MW on DC. Production of the two designs overlapped, with a combined peak output of one locomotive per week.

Both locomotives taxed the ingenuity of the engineers who designed and built them. While the shuttle locomotives had a uniform duty to perform and only one power supply voltage, many special features were necessary to integrate them into a trainset and meet stringent safety requirements. For example, the cab has to double as an office for the train captain (chef du train), who has an elaborate communications console at his disposal including monitors for the closed-circuit television security system in the passenger-vehicle shuttle vehicles.

The 31 Eurostar trains have ownership split among BR (11), SNCF (16) and SNCB (4). They have a '2 + 18' formation, consisting of two identical halves. An additional seven trains (with four fewer coaches) have been ordered by BR for services north of London. The Eurostar trains are built in France, Belgium and the UK by the Trans-Manche Super-Train Group, a consortium led by GEC Alsthom.

Eurostar trains run on three separate power supply systems, and are the first international trains to combine third-rail and catenary voltages. At first the power output will be restricted by the third-rail operation on the London–Folkestone route, but this will be rectified by the future direct London–Channel Tunnel Rail Link, due to be completed in the year 2002–3.

SHUTTLE LOCOMOTIVES

The wheel arrangement of shuttle locomotives was an important factor in the design. Topographical constraints on the location of the UK terminal mean that shuttle trains approaching it must traverse a 280m minimum radius loop of more than 180° to enter the loading and unloading platforms. With the locomotives negotiating this curve at 60km/h on every journey, minimising

Figure 1. Shuttle locomotive in the Tunnel at the UK crossover. (Source: QA Photos)

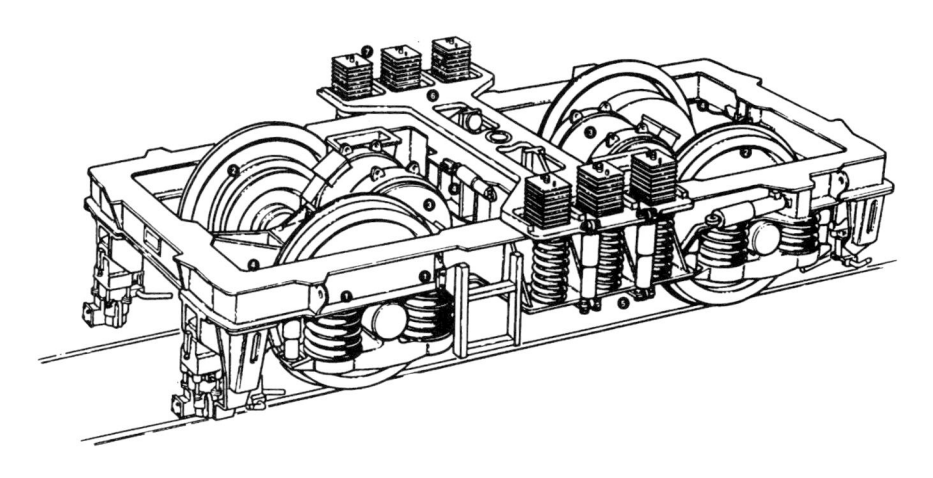

Figure 2. Shuttle locomotive bogie. 1, primary suspension and axleboxes; 2, wheels and axles; 3, traction motors; 4, traction motor links to the bogie frame; 5, secondary flexicoil suspension; 6, swinging bolster; 7, resilient suspension between bolster and body

wheel and rail wear is a major consideration. At the same time, safety and reliability for tunnel operation required the triplication of traction and braking systems. Taking all these considerations into account, a Bo-Bo-Bo configuration, with three short-wheelbase, two-axle bogies, had clear advantages over a more conventional Co-Co six-axle layout. A shuttle locomotive is shown in Figure 1.

The order for 40 locomotives (later reduced to 38) was placed in July 1989 with Brush and ABB as members of the Euroshuttle Locomotive Consortium. Both builders had experience with tri-Bo designs, Brush with its Class 30s for the New Zealand Railways' North Island electrification and ABB with Swiss Federal Railways' Class Re 6/6. Eurotunnel was able to assess the performance of the Swiss design under extended tunnel operation on the Simplon route. The shuttle locomotive bogies are based on the NZR design, with minor improvements (Figure 2). Of conventional fabricated steel construction, the bogie frame was the first Brush design to be subjected to a full UIC fatigue test.

Traction motor and gearbox are both bogie frame-mounted to minimise unsprung masses. The gearboxes are a close derivative of those used on German Federal Railways' Class 120 Bo-Bos. The single-stage parallel drive with a herringbone gear eliminates end loadings on the gearbox. The drive is taken to one wheel of each axle by a quill tube and flexible coupling.

Compared with the DC motor of the narrow-gauge NZR locomotives, the shuttle's asynchronous motor is more compact. This allows the low-level links, which transmit traction and braking forces from the bogie to the superstructure, to be connected directly to the bogie frame rather than to the traction motor casing. Each bogie has two pretensioned links running longitudinally from the centre of the bogie at about 200mm above rail level, which reduces the transfer of weight between the axles on the bogie under traction or braking.

Primary suspension is by coil springs at each axlebox, with vertical movement provided by a guide post incorporating a rolling rubber ring which provides a degree of lateral compliance. Secondary suspension also follows the NZR design, with a combination of metal Flexicoil springs and rubber stacks, plus an intermediate steel bolster. Stops allow the centre bogie up to 200mm of sideways movement allowing it to negotiate 100m radius curves in the depot. Brush had to take into account predicted aerodynamic lateral forces from the piston relief ducts in the rail tunnels when calculating the suspension roll characteristics. Yaw dampers are fitted to allow 160km/h running to be achieved with the lower primary suspension stiffness needed for the specified curving performance.

An interesting aspect of detail design is the use of the British Rail P8 wheeltread profile, whose higher conicities (up to 0.4) will keep the wheelsets out of flange contact for a greater proportion of the duty cycle than the UIC profile originally specified.

Monocoque body

The shuttle locomotive body is a conventional stressed-skin monocoque structure, with longitudinal members at solebar and cantrail levels. To permit movement to off-site workshops on the Continent for overhaul, the locomotives are built to the UIC 505-1 structure gauge. Fabricated by Qualter Hall in Barnsley, the shells set new standards for flatness on the bodyside panels. A single sheet of steel used for each bodyside skin is pretensioned to approximately 60 tonnes before the structural inner framework is welded on.

A complication was introduced by the 1250mm diameter wheels, which are the largest ever supplied by ABB British Wheelset. These leave little clearance beneath the superstructure, so all secondary longitudinal structural members must run above floor level.

Special attention was given to the bodyshell end design, as a locomotive derailed in the Tunnel can be lifted only from the buffer beams and, in extreme conditions, at the diagonal corners. This introduces high stress levels across the cab doors which represent a discontinuity in the monocoque. Brush's strain-gauge testing of the shuttle locomotive shell simulated 12 load cases including diagonal lifting.

To reduce peak collision forces, a honeycomb energy absorption capsule is installed at the cab end. Although the cab is designed to UIC end-load requirements, the energy absorption feature increases the forces transmitted into the structure in the event of a collision, and this had to be accommodated in the shell design.

The locomotives are coupled to the shuttle wagons by autocouplers which incorporate a special Scharfenberg coupler that can be operated remotely, and BSI draftgear. The Fabeg electrical heads provide connections for all control signals and the 1.5kV DC train auxiliary supply. A UIC 1.5 MN drawhook and buffers are provided at the outer end of the locomotives for emergency use.

Sealed cabs

To prevent aerodynamic pressure pulses from being transmitted into the main cab during tunnel operation, it is fabricated as a sealed pressure vessel, with the air conditioning acting as a positive displacement pump. Integrity of the pressure vessel is maintained by fitting the external cab doors with inflatable aircraft-style seals.

A one-piece fibre-reinforced phenolic moulding, using technology developed with British Petroleum, provides the streamlined nose cone. The moulds incorporate jigging facilities to ensure complete interchangeability of components between locomotives.

The main cab contains a driver's desk on the left and a train captain's position to the right, occupied at the rear of the passenger shuttles. Freight shuttles are managed from a console in the amenity vehicle.

The equipment providing the interface between the train captain's console and the shuttle vehicles was supplied by the wagon contractor for installation in the locomotive. The console contains a switch panel for the carrier and loader vehicle systems and two video monitors with split screens to display the vehicle interiors, as well as radio and telephone equipment. Touch screens allow the train captain to initiate recorded announcements in English and French.

Electrical equipment

The shuttle locomotive's main traction equipment was the responsibility of ABB Transportation Systems. ABB supplied the transformers, traction and auxiliary power supply converters and rectifiers, together with all control electronics. Brush manufactured the traction motors to an ABB design. The specified rating is based on two locomotives hauling a maximum train load of 2100 tonnes with a 30 min terminal-to-terminal journey time. Onto this are added specifications to meet three emergency cases:

- In the event of one bogie of one locomotive in a shuttle pair being isolated, the train must be able to complete the journey to schedule.
- Where one locomotive has failed totally, the train must be able to restart from a standstill and leave the Tunnel at reduced speed.
- One shuttle train must be able to rescue another which has suffered total power failure and haul or propel it out of the Tunnel.

Electric braking is regenerative only. Apart from the problem of fitting rheostatic brake resistors and their associated blowers into such a tightly packaged locomotive, the discharge of hot air into the tunnel environment was considered undesirable.

Power equipment is distributed along both sides of a central corridor in the locomotive body, with the main electronics cubicle to one side of the vestibule at the outer end. The requirement for a central corridor, to give easy access between the cab and train, imposed a particular constraint on the main transformer which had to be tall and narrow.

To avoid electromagnetic interference, communications links between the control electronics and the equipment within the locomotive are provided by an optical fibre databus in a star arrangement. The two locomotives on each train communicate over an ABB MICAS databus carried through the intervening wagons and couplers. In addition to the optical data links, each locomotive incorporates 29.5km of conventional wiring.

The 25kV 50Hz traction current is collected by one of two pantographs on each locomotive, feeding through roof-mounted high-voltage circuit breakers. The single transformer, rated at 7MVA, feeds three traction converters, mounted along the opposite side of the central corridor. The converters, each of which powers the two motors on one bogie, are completely independent. Each incorporates the now-conventional combination of four-quadrant controller (input rectifier), DC link and three-phase inverter.

Availability of gate turn-off (GTO) thyristors with a blocking voltage of 4.5kV and a turn-off current of 2.5kA has allowed the use of a DC link with three poles instead of two. This brings benefits in power rating, reduced component count, decreased harmonic interference and less pulsation in traction motor torque.

A standard triple thyristor assembly is common to both the four-quadrant controllers and the drive inverters. The thyristors are oil cooled, with each assembly making up a separate tank connected to the cooling circuit by quick-release connectors. Each oil-cooled converter is grouped with its associated heat exchanger, battery charger and resonant capacitor. Also mounted on this side of the locomotive is the brake control panel.

Installed on the same side of the central corridor as the transformer is the 750kVA auxiliary rectifier which supplies the shuttle wagon systems at 1.5kV DC. Fed by a dedicated winding on the transformer, this rectifier is a noncontrolled diode bridge.

While passing through neutral sections, each of the two locomotives will have to provide auxiliary power for all the intermediate shuttle vehicles. To allow this, the auxiliary supply is connected end-to-end through the train, with provision for temporary load shedding. The approach of neutral sections is signalled by data transmission loops on the track.

The traction motor is derived from the design used on Schweizerische Bundesbahn's Class 450 and 460 Bo-Bos. Each motor is continuously rated at 960kW at 1100 rev/min, and 56Hz, and weighs 2150kg.

Fire protection

Fire safety and resistance of materials and equipment is of the highest priority in the shuttle locomotive design. For example, the locomotive's 'nose' is a moulding made from fibreglass and Cellobond FRP resins manufactured by BP Chemicals, which meets the Tunnel's fire standards. There are special fire-resistant main power cables, contained in double-level trunking below the floor; their increased stiffness compared with those used in conventional locomotives required a more complex design.

The shuttle locomotive interior is divided into four zones, each of which has its own heat and fire detection systems. One of these systems comprises pressurised plastic tubing threaded through the various equipment cases. With the overheating of any item of equipment, the plastic tubing melts and the air pressure is released, which activates an alarm in the cab, shutting off power. A Halon 1301 discharge is triggered if necessary to extinguish any fire (after the ventilation fans have been allowed to stop). Firefighting control equipment and small Halon bottles are installed near the two cabs, with larger Halon bottles on the roof. The fire alarm system in the leading locomotive is connected to the display in the driver's cab in the rear locomotive by dedicated hardwired links that run the length of the train. In lockable cupboards on the shuttle exterior are external manual controls for the extinguishers.

CLASS 92 LOCOMOTIVES

For the 46 Class 92 locomotives that will haul passenger and freight trains operated by BR and SNCF through the Tunnel, the critical factor is the ability to operate on the 750V DC third-rail power supply in southern England, and

Figure 3. Class 92 locomotive at Dollands Moor sidings prior to October 1994 Tunnel trials. (Source: QA Photos)

their ability to avoid generating currents at frequencies that might interfere with the AC track circuits used with DC electrification. This resulted in a significant delay to their introduction, and they were not put into service until November 1994..

In 1989, BR approached 16 European and Japanese manufacturers with a specification for a dual-voltage locomotive to haul fast freight trains through the Tunnel. The locomotives, designated Class 92, would be operated by Railfreight Distribution between freight terminals from as far north of London as Glasgow, and Fréthun in France, where SNCF would take over. The specification required a maximum speed of 140km/h. Traction performance was determined by Channel Tunnel emergency operation requirements and haulage capabilities on the steepest inclines in Britain. In the Tunnel, a Class 92 has to be able to haul a train plus a dead locomotive up the 1 in 90 ruling gradient from km 23 to the British portal at 30km/h. On domestic lines in the UK, a single locomotive is required to haul a 1600-tonne train northbound up Beattock bank, a gradient of 1 in 75, from a standing start.

BR's subsidiary, European Passenger Services, also needed a small fleet of electric locomotives to haul overnight trains to and from Glasgow/London and Fréthun. To simplify procurement, and reduce unit costs, BR decided to specify the Class 92 for both duties.

Brush Traction was awarded an initial contract for 20 Class 92 locomotives in June 1990. This was followed by three further orders: ten more for Railfreight Distribution, seven for European Passenger Services and nine for SNCF. However, the locomotives are in all respects identical and can be interchanged between passenger and freight duties. Figure 3 shows a Class 92 locomotive.

Weight and space critical

Structurally, the Class 92 is derived from the bodyshell for BR's Class 60 diesel electric freight locomotive. While the structure does not have to accommodate a 24.5 tonne diesel engine, the 750V DC operating requirement ensured that the Class 92 is equally weight and space sensitive, which led to the employment of a full-time weights engineer during the design phase.

Compared with the Class 60, design of the stressed skin monocoque bodyshell was simplified by the absence of cutouts for radiator air intakes. Offsetting this was the requirement for the locomotive to be lifted at the buffer beams in the event of a derailment in the Tunnel. As with the shuttle locomotive, this imposes additional stresses across the cab door area.

Subcontractor for the bodyshells was Bombardier Prorail of Wakefield. A notable innovation is the use of a structural one-piece steel pressing for the cab end. In recent years, glass reinforced polyester or phenolic composite cab ends have become almost the industry standard. However, Brush had found that variations in hand laid-up composite mouldings could cause alignment problems when fitting to a rigid steel shell.

Investigation showed that modern motor industry techniques had made a one-piece steel pressing a cheaper and dimensionally more consistent alternative to composite cab ends. In addition, the steel could be welded into the structure, adding strength and eliminating the unsightly line of flexible filler needed between the composite cab and bodyshell.

A second one-piece steel pressing provides the outer skin of the cab roof. The cab air-conditioning module is mounted in the locomotive roof behind the cab. As with the shuttle locomotives, body side panels are prestressed before welding to minimise ripple.

Also derived from the Class 60 is the bogie design. The three-axle bogie has a fabricated steel frame with coil spring primary suspension and axleboxes located by guide posts with rolling rubber rings. Flexicoil secondary suspension provides the necessary ride quality. Traction motors are axle hung, with pinion and gear direct drive to the axle.

Loco 2000 parentage

Electrically, the ABB traction equipment is derived from the Swiss Loco 2000 design, with a four-quadrant converter incorporating GTO thyristors feeding inverters through a high-voltage DC link. The locomotive's continuous rating is 5000kW on 25kV 50Hz and 4000kW on 750V DC. When operating on the third rail, the incoming 750V supply must be increased to the nominal DC link voltage of 2.8kV. This is achieved by reconfiguring the four-quadrant controller to work as a step-up DC chopper.

Nominal maximum tractive effort is 360kN, but a starting boost to 400kN is available. This is equivalent to 200kN tractive effort from a single bogie and, when a single locomotive is hauling a train, the failure of one traction package automatically enables the remaining package to provide a tractive effort of 200kN. This will be sufficient to restart a 1300-tonne train on the 1 in 90 ruling gradient in the Tunnel.

Restriction of the nominal tractive effort to 360kN reflects the trailing load limitations imposed by the UIC couplings used on international freight vehicles. For this reason, the microprocessor-based control system incorporates a facility which limits the combined tractive effort automatically if two Class 92s are coupled together in multiple.

For both close-coupled multiple operation and when running with a locomotive at each end of the train, control signals are transmitted between the locomotives over a two-wire time division multiplex system. However, certain safety critical control functions are hardwired.

Although the Class 92 has hardware capable of providing regenerative braking into both DC and AC supplies, software is not loaded for DC regeneration, and AC regenerative braking will be restricted to the Tunnel. On BR, the locomotive is configured for rheostatic braking only. In this mode the braking resistors are connected across the DC link. Electric braking is rated at 5000kW in both cases.

Channel Tunnel safety policy is based on a train completing its journey in the event of equipment failure. Initially, this required each train to be powered by at least two locomotives, but single-heading of trains up to 1300 tonnes has been accepted. The need for high reliability with a single locomotive led BR to specify that the Class 92 traction drive should be arranged to create what is effectively two locomotives in one.

To achieve this, each bogie group of three motors has its own independent drive, and the locomotive superstructure is divided into three compartments.

Each end compartment houses the traction equipment for one bogie group with the central compartment containing the electronic controls.

Fire-resistant bulkheads with a 30 min fire rating separate the compartments which have independently operated fire suppression equipment. Control cabling between the central compartment and the traction equipment runs in trunking with a similar fire rating.

Current collection

Complementing the independent drives, other equipment has been duplicated where possible. Thus there are two 25kV pantographs. Current collection for a high-power locomotive on third rail is dominated by the effect of gaps in the contact rail. While the design philosophy is based on each bogie and its drive operating independently, the need to provide as long a span as possible for the contact shoes has made interbogie cabling necessary.

Because of the asymmetric axle spacing, each bogie has three sets of retractable shoegear on each side, installed as a pair of shoes plus a single shoe. Both pairs of shoes at the outer end of each bogie are cross-connected to the two single shoes on the inner end of the other bogie. Even so, such a high-power locomotive is more susceptible to arcing when crossing gaps. Because of this, the control software in the Class 92's microprocessor is programmed to detect arcing when the leading bogie crosses a gap and cuts back power automatically to reduce arcing as the trailing bogie reaches the same point.

Highlighting the design challenge posed by dual-voltage operation, space had to be found on each bogie for these current collection cables plus AC and DC earth return cables, in addition to the traction motor power feeds.

Current from the pantographs passes through individual vacuum circuit breakers to the transformer mounted below the underframe between the bogies. This oil-cooled unit shares a tank with the inductors for the DC step-up chopper.

Windings and inductors are arranged to provide independent circuits for each drive. With a maximum direct current of 6800A, these inductors are substantial components. The tank is also partially double skinned to meet safety requirements which specify resistance to penetration. As a result, the complete assembly weighs nearly 17 tonnes. When installed and bolted to the solebar, the top of the transformer tank forms the floor of the central control equipment compartment. Flexible busbars take power to cabling in the solebars. To simplify assembly, the adjacent bulkheads are bolted, rather than welded, into position and are located after the transformer and its associated services have been installed.

Equipment layout on each side of a central corridor is identical for the two end compartments, but laterally transposed. Working forward from the bulkhead with the central compartment, to the left is the main cooler group, the four-quadrant controller, the inverter and the resonant capacitors of the traction drive. The inverter feeds the three traction motors on the bogie at that end, with the motors wired permanently in parallel.

To the right of the corridor are the braking resistor, DC equipment/auxiliary power supply cubicle, traction motor blower, body ventilation fan, the main compressed air reservoir and the Halon storage bottles for the fire extinguisher system.

Each asynchronous traction motor is rated at 840kW at 981 rev/min. Although similar to those fitted to the shuttle locomotives, the stator has heavier gauge windings, reflecting the greater emphasis on low speed/high tractive effort operation with the Class 92, and mechanical strength has been increased significantly to withstand the axle-hung environment.

The central compartment houses the brake equipment frame and the electronics for locomotive control, TVM430 automatic train protection and firefighting. TVM430 is required only for service through the Tunnel. In the UK, and on SNCF tracks between the Tunnel and Fréthun, Class 92 will rely on BR's automatic warning system.

Control commands within the locomotive are transmitted over the standard ABB MICAS 2 databus. A separate MICAS databus is provided for communication between locomotives working in multiple at each end of the train.

Design of the auxiliary power supply is complicated by the dual-voltage operation and the hotel power required for night stock which will have an electrical load of up to 1MW. As with the propulsion equipment, systems are duplicated.

Auxiliaries

Each traction package has an auxiliary power supply tapping on the DC link. Taking the auxiliary supply from the DC link is an unusual feature which was necessary to provide a common source of auxiliary power when operating on DC or AC systems. A chopper steps down the DC link voltage to 570V which is used to feed three auxiliary inverters supplying the following items for each traction package:

- Traction motor blower and radiator cooling fan; this is a variable-frequency supply so that cooling power can be matched to demand, reducing fan noise in stations.
- Oil circulation pumps, air compressor, battery charger and cab air conditioning; this is a fixed-frequency supply.
- Brake resistor cooling fan, also requiring a variable-frequency supply.

Auxiliary power for passenger trains is supplied by a separate winding on the transformer when operating on 25kV routes. This provides the standard 893V AC train line supply which is rectified on the vehicles. On DC lines, the raw 750V supply is fed down the train line.

Cooling air flows

With such a tightly packaged and highly rated locomotive, air flow management is a key area of design. Each bogie group has its own cooler group and air management system. Air for the cooler group is drawn in at roof level and passes through two radiators in series before being exhausted through the floor. The upper radiator cools the converter oil circuit, while the lower one provides the cooling for the half of the transformer supplying the bogie group.

Air for brake resistor cooling is drawn in from under the floor. After passing up a stainless steel trunking, it is exhausted at roof level at temperatures of up to 400°C. Deflector vanes are fitted to prevent this exhaust air from being ingested by the cooler group intake.

Axial fans are used for the traction motor blowers. The high-efficiency blowers are supplied by A K Fans and their compact size simplifies installation in the restricted space. From the blower, the air passes via a splitter box to the trunking feeding the three traction motors. Sharing a common intake with the traction motor blower is the body ventilation fan. This blows air into the central corridor, to be exhausted at the cooler group.

Air flow over the heat sinks of the auxiliary inverters is provided by tapping the low pressure intake of the traction motor blower so that air is sucked into the inverter casing from the corridor. A similar arrangement, using the low pressure trunking of the cooler group intake, is used to provide low velocity air flow over the main traction drive inverter electronics.

In contrast to the drive compartments, the central area has a single roof-level ventilation fan drawing in air through a water separator.

Fire protection

Eurotunnel has the most stringent fire regulations in Europe and these have been reflected in the design of the Class 92. Brush included fire performance in the materials control process for the locomotive, based on BS6853 and BR fire specification CP/DDE/101. All nonmetallic materials and components used in the Class 92 were fire rated. Where a supplier could not provide the necessary certification, products ranging from plastic push buttons to the driver's seat were fire tested.

Figure 4. Eurostar trainset

Kidde Fire Protection is supplying the zoned fire extinguishing system for the Class 92. This provides independent and controllable detection and extinguisher systems for the three compartments. The complete system is monitored by a central microprocessor-based unit. To ensure reliable protection in the event of fire, the bottles containing the Halon for each end compartment are installed at the opposite end.

In the cab is a pair of buttons for each of the three zones. In the event of a fire, the relevant button flashes. The driver then has the option of letting the automatic extinguisher operate or pressing the button to delay initiation of the extinguisher in that zone. The delay is repeated each time the button is pressed. If the delay button is not pressed, there is a 30s pause while the fans and blowers in the zone are run down before the Halon is released. If the fire is not suppressed, the driver has the option of a further 'one shot' button which floods all the zones with Halon.

EUROSTAR TRAINS

The design of the Eurostar trainset (Figure 4) is broadly based on that of the French TGV (*train à grande vitesse*), with a number of major changes, chiefly concerning the loading gauge and the ability to run on three separate supply voltages (see later). A major change was the adoption of a British-designed asynchronous drive in place of TGVs' synchronous motors. The profile of the TGVs below platform level is slightly too large to permit their operation in the UK, so the Eurostar trains are narrower, necessitating alterations to the bogies' air-suspension and the coach width. There are retractable footboards at the doorways extending to varying height/widths, enabling boarding and alighting from UK and Continental platforms, and the walkways inside the Tunnel.

Eurostar power cars and vehicles were built in the UK, France and Belgium by a consortium of manufacturers led by GEC Alsthom. GEC Alsthom in the UK led on the electric power equipment and GEC Alsthom in France on the mechanical portion.

Fire safety design features specific to the Tunnel have been included in these trains. Fire detection and extinguishing equipment is incorporated in the traction compartments, fire-resistant materials have been used wherever possible and the doors at the end of each passenger saloon, and the floors, are 30 min fire resistant.

The Eurostar trains, operated by a single driver, use the same TVM430 cab-signalling and speed-control system as the Channel Tunnel and the TGV Nord Europe line, plus the conventional cab-warning systems used on the national railway networks. For the journey through the Tunnel, the Eurostar trains operate under Eurotunnel's railway control centre.

The design and construction materials of the TGV sets had to undergo many changes so that Eurostar trains could meet the safety requirements of the Tunnel. The driver or train captain has to be able to divide the train in the Tunnel (as with the passenger-vehicle shuttles), which can be done in three ways: uncoupling either of the power cars, or splitting the train between the two centre coaches. So that Eurostar trains can be hauled out of the Tunnel if the power supply fails, the couplings, brakings and services connections are compatible with Eurotunnel's diesel-electric locomotives.

Eurostar trains need to be able to run on three separate supply systems: the Paris-Tunnel SNCF lines, the Belgian high-speed line and the Tunnel itself (25kV AC); the existing Belgian line (3000V DC); and Folkestone to London in the UK (third-rail system, nominal 750V DC). In the third case, the need to provide retractable shoegear on the power bogies to pick up current from the third rail, and the equipment to use the low-voltage supply, has complicated the design of the Eurostar trains.

Because of third-rail voltage drops, and the restricted BR Folkestone–London supply arrangements, the Eurostar trains' power output is restricted in this stretch to less than a third of its potential in the Tunnel and France. However, with the completion of the London–Channel Tunnel Rail Link incorporating a 25kV catenary, Eurostar trains will be able to perform to the full extent of their capabilities.

13

Passenger-vehicle shuttle fleet

PETER SEMMENS AND
YVES MACHEFERT-TASSIN

Each passenger-vehicle shuttle consists of two separate halves, or rakes, single deck and double deck. Each half contains two loading/unloading wagons and 12 carrier wagons. All shuttles have a locomotive at each end.

Eurotunnel's initial order was for nine passenger-vehicle shuttles. The total fleet is therefore:

- 18 single-deck loading/unloading wagons plus one spare
- 18 double-deck loading/unloading wagons plus one spare
- 108 single-deck wagons
- 108 double-deck wagons.

WAGON DESIGN

The wagons that form the shuttle trains are the largest railway vehicles in the world. They occupy over half the cross-sectional area of the tunnel, where they travel for 50km at speeds of up to 140km/h. Such speeds cause rapid changes in the external pressure, hence the need for sophisticated systems and equipment to prevent these fluctuations making the journey uncomfortable for passengers and to ensure their safety.

Each passenger-vehicle shuttle train is 776m long, and consists of 28 wagons and two locomotives. The passenger-vehicle shuttle wagons were built by the Euroshuttle Wagons Consortium (ESCW), composed of GMT-Bombardier (Canada), La Brugeoise et Nivelles (BN) (Belgium) and ANF Industrie (France). These are all now subsidiaries of Bombardier Eurorail, which subcontracted the construction of the 508 bogies required to GEC Alsthom at its Le Creusot works.

The passenger-vehicle shuttles, usually known as tourist shuttles, have two types of wagon. The double-deck wagon is designed for cars with a maximum height of 1.85m (Figure 1). The single-deck wagons are used by cars with high trailers or roof racks, coaches, minibuses, and other larger vehicles more than 1.85m high. Both types are 26m long, with a height of 5.6m above rail level. They are designed for a maximum speed of 160km/h, although initially they

Figure 1. Passenger-vehicle shuttle: this double-deck carrier wagon, on delivery in France in July 1993, is not yet attached to its bogies. (Source: QA Photos)

will operate at 140km/h. Each single-deck wagon weighs 63 tonnes empty: double-deck wagons are 2–4 tonnes heavier (depending on design, see later) because of the extra floor.

The wagons are made of stainless steel. All their structures and fittings are built to provide the best insulation against noise and thermal variations during their operational life. Insulation, in particular with fire-resistant rockwool, ensures a noise level inside the wagon of less than 80dB.

CARRIER WAGONS

Each single-deck wagon can carry any combination of vehicles weighing up to a total of 24 tonnes. Each deck in a double-deck wagon is designed to support a total load of up to 12 tonnes; this is sufficient to take five conventional cars. For all carrier wagons, the maximum rail-axle load is 22 tonnes. Figures 2 and 3 show the interiors of the single-deck and double-deck wagons, respectively.

Two slightly different designs of single- and double-deck wagons, A and S, have been built. These are coupled semipermanently into triplets in the formation A–S–A. From the user's point of view, the only difference is that double-deck S wagons have a passenger stairway in the middle and toilets on both decks. There is no space for toilets in single-deck carrier wagons; they are located instead in the loading/unloading wagons. The coupling arrangements of each design also differ.

When a train is being marshalled for service, triplets can be added or removed by ordinary shunting movements, normally using the automatic low-level couplers on their outer ends. However, because the individual wagons within a triplet are joined by a semipermanent bar coupling, they can only be separated and joined by maintenance staff.

When road vehicles enter the shuttles they drive along a continuous flat floor; side kerbs, 92mm high on double-deck and 150mm high on single-deck

Figure 2. Passenger-vehicle shuttle: interior of single-deck carrier wagon. (Source: QA Photos)

wagons, prevent them approaching the walls too closely. Floor plates bridge the gap between each carriage just above the couplers, enabling vehicles to drive along the half-shuttle from one end to the other during loading and unloading operations. There are separate routes through on each level of the double-deck wagon. Airtight bellows enclose all the space between the wagons, including connecting bridges and couplers, and are designed to ensure that sudden changes of air pressure in the Tunnel do not affect passengers; their size and fire-resistance capabilities make them unique in the railway world.

The door systems that close off the wagon ends (Figure 4) consist of two side sections hinged from the wagon side with a descending roller shutter between them; they were made by De Dietrich in France. When these are in place and the seals between them have been inflated, the whole door assembly becomes a barrier capable of resisting a fire inside the wagon for up to 30 minutes, long

Figure 3. Passenger-vehicle shuttle: interior of double-deck carrier wagons. (Source: QA Photos)

enough for the shuttles to reach either end of the Tunnel. Passengers may use pass doors in each hinged section to reach the toilets in the central S wagon of each double-deck triplet, or those at the ends of the single-deck loading wagons. The opening of these pass doors is hand operated, using a push-button with a two-second delay.

Air-operated steel protection arms, rising out of the floor near the ends of the wagons, prevent vehicles colliding with the fire doors should the handbrake on the car nearest a door be insufficiently applied or accidentally released. Le Shuttle staff guide each vehicle into position and ensure that the handbrake is on. When the wagon is full, they close the doors and shutters between wagons, and raise the steel protection arms.

Normally, passengers drive on and off the shuttles in their vehicles. Two emergency doors are provided on the walkways on each side of each wagon for use should it ever be necessary to evacuate a shuttle train in the Tunnel. These doors are normally locked, but can be unlocked pneumatically by the train captain (chef de train) and also locally if the shuttle speed is below 5km/h. If the normal mechanism cannot be operated, manual opening is also possible in an emergency if the shuttle has stopped. Passengers would alight on to the continuous walkway on the side nearest the service tunnel.

STABILISERS

When heavy vehicles drive on, along or off the shuttles, during loading and unloading in the terminals, they could cause unacceptably large deflections in the wagons' suspension system as their centre of gravity moves. To prevent this happening, stabilising beams or jacks are provided under those wagons that have to handle heavy loads.

Figure 4. Passenger-vehicle-shuttle: fire doors at end of double-deck carrier wagon. Note that bellows enclose couplings. (Source: QA Photos)

Stabilising beams are fitted to freight and to single-deck passenger-vehicle carrier wagons to resist the variable forces caused by heavy vehicles driving along inside the shuttles. There are two beams per wagon, each operated by two hydraulic pistons. The beams move down to bear on the two rails of the track, and lift the wagon slightly off its suspension.

Stabilising jacks are provided on freight and single-deck passenger loading/unloading wagons to resist the tilting and transverse forces caused by heavy vehicles loading. The possible forces are too great for the track to resist, so concrete beams have been cast alongside the track. There are six jacks, of two different types, which locate on these beams and lift the wagon slightly off its suspension.

There are no beams or jacks on the double-deck shuttles, as the loads imposed by cars are much less.

LOADING/UNLOADING WAGONS

A loading/unloading wagon is positioned at each end of each half-shuttle. Vehicles drive in through the rear loading wagon and drive out through the front unloading wagon. The double- and single-deck loading wagons differ considerably in design.

Double-deck loading wagons

Externally, double-deck loading wagons closely resemble carrier wagons, although the window arrangements differ. Two large sliding doors, 6m long, on each side let vehicles drive in and out. Cars entering and leaving the lower deck use the doors nearer the carrier wagons. The further doors lead to an internal ramp, for cars to reach or leave the upper deck. Normally, the two decks load and unload from adjacent platforms simultaneously, thus speeding up the operation. To span the space between doors and platforms, two bridging plates can lowered from inside the wagon; these are operated by hydraulic rams concealed in the floor.

A small seating area for motorcyclists and their passengers is positioned over the vehicles' entrance at the outer end of each loading/unloading wagon. The motorcycles travel securely housed under the ramp in a special cubicle, which accommodates three motorcycles in each wagon. A toilet is provided alongside the seating area. Spare wheelchairs for disabled people are also located inside the motorcycle compartment.

A small driving cab positioned in one corner of the outer end of the loading wagons is used for moving groups of carrier wagons in the stabling and maintenance areas. The cab, which is normally locked, contains controls for low-speed driving and a panel for operating and monitoring the wagon's own systems, such as the loading doors; it is not fitted with a cab signalling system. A control box can also be connected outside the wagon to enable the maintenance staff to move a half-shuttle at a speed of no more than 3.5km/h. A helical staircase alongside the cab connects the two decks.

Single-deck loading wagons

Single-deck shuttles require a more complex design of loading wagon capable of taking long and heavy vehicles. The function of the single-deck loading wagon is to provide sufficient lateral space to allow a coach, or a car with a caravan, to align itself to drive through the rake of carrier wagons. The loading wagon does this by forming a bridge between each pair of terminal platforms. The bodywork, whose sole purpose is to link the carrier wagons with the toilets, consists of three hood canopies that telescope onto a fixed canopy at the outer end of the wagon during loading and unloading, so freeing three-quarters of the wagon's length for vehicles to manoeuvre. Long loading plates on each side hinge down to bridge the gap between the wagon and the platform, thus allowing vehicles to use the entire width of the loading wagon and both platforms to align themselves. When every vehicle has loaded, the loading plates are lifted and the canopies moved to enclose the entire floor area; seals between them are inflated to reduce the effects of pressure changes

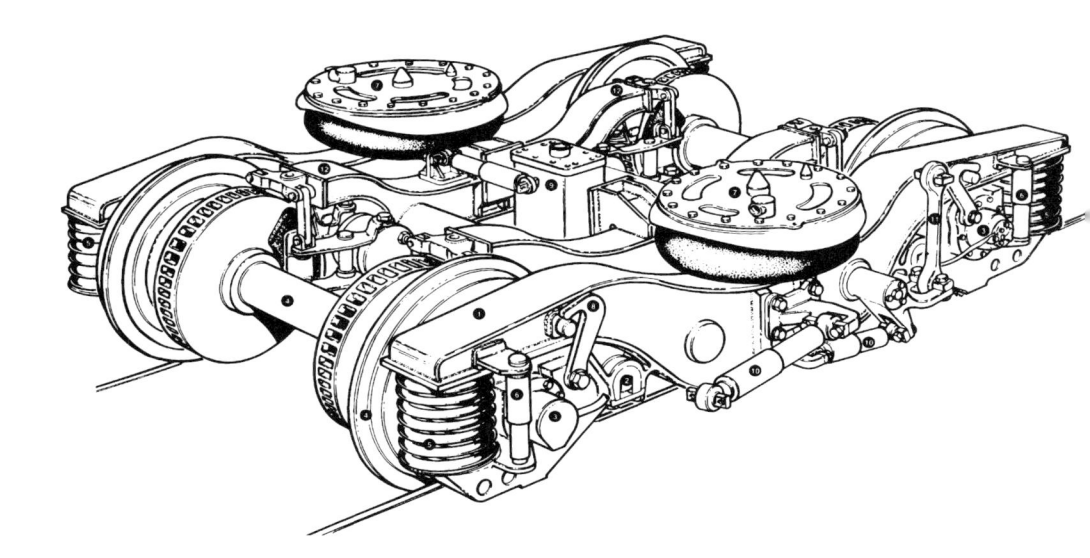

Figure 5. Passenger-vehicle shuttle: bogie. (Source: Eurotunnel)

in the Tunnel. The canopies are moved by wire ropes and hydraulic motors; hand operation is also possible.

The greater width of vehicles using the single-deck shuttles means there is no room for toilets in the carrier wagons. Instead, they are situated in an air-conditioned amenity area at the end of each loading wagon, behind a small driving cab, which has the same functions as the cab in the double-deck loading wagon.

BOGIES

The bogies must provide a satisfactory ride for passengers whether the wagon is fully loaded or nearly empty. They must also ensure that the wagon remains within the loading gauge at all times.

As shown in Figure 5, the bogie has a fabricated frame, H-shaped in plan (1). Under the upswept ends, pivots for the short radius arms (2) support the axleboxes (3) for the wheelsets (4). These rise and fall with 90mm vertical play to accommodate any irregularities in the track, and the coil springs (5) provide the primary suspension. Hydraulic dampers (6) are used to prevent any bounce. The whole bogie is located on the underframe of the wagon by the central pivot and attached by the secondary suspension, which consists of two large air bags (7). This arrangement is widely used on passenger rolling stock, including the latest TGV and Eurostar trains, although the empty/

full weight changes on the Eurotunnel stock are far greater than those on conventional trains.

As well as giving the smoothest and quietest ride for passengers, the air bags ensure that the bogie remains centred on curves. They also provide part of the restoring forces to keep the bogie parallel with the vehicles. The air bags are inflated from the train's compressed-air system, through a levelling valve actuated by the height of the body above the rail. Any change in the overall weight of the wagon causes it to rise or fall on its springs; the levelling valve adjusts the air pressure in the bags to restore the vehicle to its normal level. The primary suspension does not need any similar compensating device, as the top of the bogie frame remains between 812 and 815mm above rail level, whatever the load. Hooks (8), attached to the axleboxes, prevent the bogie frame rising too high in the unloaded condition, or when the wagons are jacked up for maintenance.

The pivot (9) is rigidly fixed to the underframe of the wagon, and a small amount of lateral play is permitted by the Z-linkage arrangement. To damp any oscillations and to prevent hunting (rapid side-to-side angular motion) of the bogie at speed, two sets of hydraulic dampers (10) are provided between the bogie frame and the vehicle body. The centre of gravity of a loaded wagon, especially the double-deck version, could lead to a roll developing during travel. To prevent this, a torsion bar passes through the bogie frame tubes, vertical links (11) connecting the cranks on its ends with the two sides of the body. Any attempt by the body to roll is thus opposed by the links twisting the ends of the torsion bar in opposite directions.

Four additional short arms (12) are attached to the cross-member of the bogie frame to support the braking equipment. (For clarity, this is not shown in the diagram, but the corresponding arrangements can be seen in the diagram of the freight shuttle bogie in Chapter 14.) However the brake discs on which the brake pads engage can be seen, rigidly fixed to the axles between the wheels. On the passenger wagons the air bag suspension pressure is also linked to the brake control system. Since the air pressure in the bags is related to the load of the vehicle, it is also used to vary the braking effort to match this load.

BRAKING SYSTEMS

Both the passenger and the freight shuttle wagons rely on compressed air for the braking operation, and this provides the basic fail-safe system. Two air pipes run the length of the train. One, the air reservoir pipe, is used to charge the reservoirs on the underframe of each wagon with compressed air from the locomotives. Normally the pressure in the air brake pipe is controlled by the driver's brake valve; this is reduced to apply the brakes throughout the train, and restored to normal pressure to release them.

On each vehicle the brake operation is controlled by a complicated valve called a distributor, which uses the changes in pressure in the brake pipe to regulate the flow of air from the reservoirs into the brake cylinders. This arrangement provides the 'automatic' safety system required by law to be fitted to every passenger train. Should the pipes between adjacent wagons be severed, the release of air pressure would immediately cause the brakes to be applied on both sections of the train. Nonreturn valves between the air reservoir pipe and the reservoirs ensure that the pressure in the reservoirs remains available to activate the brake cylinders.

AIR CONDITIONING

For several decades air conditioning has been provided on high-quality railways throughout the world. This has markedly improved the ambience of train travel, cooling the train in hot weather and reducing draughts and noise from outside, especially in tunnels. While air conditioning is an important feature of the passenger-vehicle wagons, the system is required to perform another important and much more unusual function. This is to purge the wagons of fumes while vehicles are being loaded and unloaded, and so avoid passengers having to endure an unpleasant atmosphere during the journey.

There follows a description of the air-conditioning system in a double-deck carriage, whose two separate car decks require a more complex system than that in single-deck carriages. Two 2-tonne air-conditioning units with a capacity of 60kW are fixed to the bottom of each wagon underframe. These heat or cool the air depending on outside conditions. The heating is provided electrically using power from the locomotives, which is fed along the 1500V 'train line' running the length of the shuttle. The cooling units are powered from the same source via static converters in the A wagons; they may require more energy during hot weather than the heaters consume in winter. The system is designed to keep the temperature in the passenger areas in the 'comfort envelope' of 18–20°C), even though the corresponding external temperatures vary from −5 to 38°C).

The treated air is circulated through ducts to lines of diffusers along both side walls of the wagons. Normally, air is extracted through grilles just above floor level, returning to the air-conditioning units to be recycled, except for approximately 15% which is purged from the system by the underfloor fan. The main fresh-air fan makes up the displaced quantity, the air entering through filters. This system ensures that the noise level inside the carriages is no louder than that on high-speed passenger trains and that any pressure surges from the piston relief ducts will not cause discomfort. During a single trip, each passenger can be sure of receiving more than 8m^3 of newly conditioned fresh air.

A further series of fans, each capable of handling 2500m^3/h, is used to purge the wagon interiors during loading and unloading. Each double-deck wagon contains eight fans, each single-deck wagon four. The fans pull the air from a series of ducts under the floors, with inlets positioned to catch the fumes from the vehicles before they spread around the wagon, and then exhaust them through grilles positioned high on the outside of the wagons. The flow is sufficiently rapid to enable all the air inside the wagons to be changed once every 70 seconds. This is supplemented by fresh air blowing in through the loading doors when these are open. This purging system is operated manually, either by the train captain or locally.

This air-conditioning system requires no fewer than 38 electric motors on each wagon. Like every other system on the train, it is monitored by the train captain's equipment, which provides a warning if anything goes wrong. Fire dampers fitted at vital points throughout the ducting systems can be closed to stop the spread of fumes or smoke.

To provide power to this complex system, two 90kW units are provided under two of each triplet of carriages. To ensure power if the train line is de-energised, batteries of 300Ah capacity at 110V for the double-deck wagons and 250Ah for the single-deck wagons are provided in the underframe.

COMMUNICATIONS

A complex audio and visual communications system enables the train captain to communicate with the passengers and crew as well as with the control centre. After passengers joining a shuttle have stopped their vehicles, they are briefed through the public information system, in both English and French, with safety and travel information. This is supplemented by text on the dot matrix screens in the ceiling ahead of them. During the 35-minute journey further announcements will be made and on arrival instructions are given to drive out of the shuttle. Throughout the journey it is also possible to tune a car radio to a special frequency to receive information about the journey.

Responsibility for the safety and operation of the train rests with the train captain, and all the audio and visual messages to the passengers originate in the train captain's 'office' in the cab of the rear locomotive. Recorded, routine information is transmitted by pressing the appropriate button or touch-screen monitor. Special messages can be broadcast either to the whole train or to particular wagons. In addition, a special call system enables passengers to contact the train captain directly.

The train captain monitors the entire train primarily by means of a closed-circuit television system, with a camera at both ends of each wagon deck. All the images from a given wagon are presented simultaneously on a split screen. In normal operating conditions each wagon is scanned in turn. However, the captain can 'home in' on a particular wagon at any time through a touch screen. In addition, if any sensor reports an important abnormal condition, the system automatically shows the wagon concerned; this largely relates to fire alarm and protection equipment.

All the other equipment on the shuttle is monitored by the train captain, who checks its performance. If any item of equipment, such as a fan in the air-conditioning system, were to stop, the captain's attention would immediately be drawn to it. Most equipment resets or reconfigures itself automatically to deal with this sort of eventuality. If human action is needed, this can be provided by the train captain, who can instruct the equipment to carry out the necessary action by calling up the system concerned on the monitor screens and pressing the appropriate button.

The couplings between the wagons carry some 200 connectors for the circuits that run the length of the train carrying information to and from the train captain's office. Some of these circuits are dedicated to a particular function, while others are shared by a number of different systems. In all cases time-division multiplex redundant circuits are used. Each constantly scans the performance of all the functions involved, and receives the answers to its electronic questions a millisecond later.

At the front of the train, the driver operates the locomotives, obeying the commands of the signalling system. The driver and train captain can discuss any unusual occurrence over the internal telephone system, and the train captain can also contact other crew members by radiotelephone. A radio voice-link to the train enables information to be passed to and from the control centre. Other recorded or coded messages can also be transmitted directly.

The communications systems on the shuttles are thus extremely comprehensive. An impression of the scale of the equipment is provided by the fact that each passenger-vehicle shuttle wagon contains over 50km of wiring, a considerable proportion of which is involved with communications.

Standby high-capacity batteries are provided to power the communication circuits.

FIRE PREVENTION

The shuttle wagons are equipped with a comprehensive fire detection and extinguishing system. Each wagon contains passenger-operated alarms as well as a series of detectors that are automatically triggered by smoke, fumes and inflammable vapours. These detectors are sensitive to ions, or ultraviolet radiations, smoke and gases, and are located in the ceiling, walls and floor or drainage duct of each wagon.

Even though smoking is not permitted anywhere on the shuttle, it is important to avoid false alarms that could cause unnecessary action by the train crew or concerns to the passengers. For the reason, two ordinary detectors or one passenger-operated alarm have to operate before the system registers a 'level 1' alarm. This alerts the train captain in the rear locomotive as well as the train crew, who immediately go to the wagon concerned to investigate and deal with the situation. The wagon's air-conditioning system is also automatically shut down and its air duct fire dampers are closed.

Two hand-operated portable extinguishers in each wagon can be used to deal with a small fire. If the smoke density increases further, the sensors set off a 'level 2' alarm, which causes the public address system to instruct passengers to evacuate the wagon concerned. Guided by the crew, they enter the adjacent wagons through the end pass doors. If the problem occurs in the part of the shuttle occupied by disabled passengers, they are given special assistance.

If the smoke density continues to increase, a 'level 3' alarm is triggered; this automatically discharges Halon gas into the area concerned. This is a very effective means of putting a fire out, and similar equipment is provided in many computer rooms throughout the world. If necessary, crew members can make a second manual Halon discharge, since each wagon has substantial reserves. Once everyone has been safely evacuated, the crew can use breathing apparatus to search the affected wagon for the cause of the fire.

Sensors also detect fuel leaking from a vehicle on to the wagon floor, and trigger a discharge of water with foaming agent to flush it into a sump below floor level. The sump has its own fire extinguishing system and is automatically emptied when the shuttle reaches the terminal maintenance and washing area.

14

Freight shuttle fleet

PETER SEMMENS AND YVES MACHEFERT-TASSIN

Each freight shuttle is formed of two separate halves, or rakes. Each half contains one loading and one unloading wagon and 14 carrier wagons. There is a locomotive at each end of the shuttle, with a club car behind the leading one.

Eurotunnel's initial order was for eight shuttles. The total fleet is therefore:

- 33 loading/unloading wagons, including one spare
- 228 carrier wagons, including four spares
- 9 club cars, including one spare.

WAGON DESIGN

Under the terms of Eurotunnel's Concession, the freight shuttle wagons have to be capable of carrying 44-tonne lorries. Constructing a railway wagon to fulfil this requirement while remaining within the 22-tonne axle-load limit imposed by the track presented the designers with a difficult task. A closed design, like that of the passenger-vehicle shuttle carriages, would have added considerably to the weight, and there was no justification in providing full passenger comfort and safety to enable a relatively small number of drivers to travel through the Tunnel in the cabs of their lorries.

Accordingly, Eurotunnel decided to carry lorries in semi-open wagons with drivers travelling in a club car at the front of the train, where they can rest away from their vehicles. Separating drivers from their vehicles did away with the need for air conditioning and many communications systems, as well as for end doors on the wagons. The semi-open design prevents anything that might work loose on a lorry blowing about and causing the catenary to arc, or damaging lineside equipment. Figure 1 shows a loading/unloading wagon and a carrier wagon.

CARRIER WAGONS

The wagon structure is a trellis of stainless steel, assembled by standard production techniques and spotwelded. The design is most unusual, with

Figure 1. Shuttle locomotive pulling freight loading/unloading wagon and freight carrier wagon out of the French portal: 1st June 1994, the first day of commercial rail freight through the Channel Tunnel. (Source: QA Photos)

diagonal members positioned where the structure is most stressed. This arrangement provides a more efficient structure than an open flat-wagon design. As on the passenger-vehicle wagons, overlapping horizontal plates provide a bridge between wagons, but the ends are otherwise entirely open.

Drivers, their passengers and Le Shuttle staff enter and leave the wagon through a door-sized opening on each side. When loading is complete, a bus picks up drivers and any passengers and drives them along the platform to the club car marshalled at the front of the shuttle, just behind the locomotive.

The total length of the freight shuttle, including its two locomotives, the club car, and the two identical halves, or rakes (each with 14 carrier wagons, semipermanently coupled, and a loading and an unloading wagon) is 730m. This is just 46m shorter than the passenger shuttle; the same platforms can therefore be used for both types of shuttle.

Lorries enter the shuttle via a loading wagon, and then drive through the carrier wagons until signalled to stop; side kerbs, 150mm high, act as guides. Although protection arms are not installed, each lorry is carefully positioned so as to balance its weight between the wagon's two bogies, and chocked to prevent it moving. A sunken section with drainage facilities runs along the centre of the wagon floor to collect any leaking oil etc.; this is automatically emptied during washing and maintenance. Overhead lighting is provided, and an interior control panel enables staff to monitor equipment mounted on the underframe.

The electrical system of the wagons is relatively simple. Supplies at 400V three-phase AC and 110V DC provide power for lighting, jacks and bridging-plate operation and for the lorry plug-in system. Lorries with temperature control or refrigeration equipment can plug into the two electrical supply points (400V AC or 110V DC) provided on each wagon side, since the diesel motor that acts as the normal power source for these purposes must be shut down during the journey.

Stabilising beams are provided on the carrier wagons, and stabilising jacks on the loading/unloading wagons, to prevent deflection in the suspension system and maintain complete stability during loading and unloading.

Breda-Fiat was responsible for manufacturing the freight shuttles, and the bogies were constructed by Fiat Ferroviaria at its Savigliano works near Turin.

LOADING/UNLOADING WAGONS

The size and weight of lorries make loading and unloading a much more complicated operation than loading cars (Figure 2). Manoeuvring space is required to align an articulated vehicle, or a lorry and trailer, so that it can drive through the train of carrier wagons. Although at first sight the freight

Figure 2. Unloading lorries from the freight shuttle: Calais Terminal (left) and Folkestone Terminal (right). (Source: QA Photos)

loading wagons appear to be simple low-sided wagons, sophisticated features have been incorporated to cope with heavy and moving loads.

The wagon's sides are in fact bridging plates which, during loading and unloading, are lowered by concealed hydraulic jacks (to the accompaniment of warning bleeps) to cover the gap between the floor of the wagon and the platforms on both sides. Stabilising jacks are also lowered onto longitudinal beams beside the tracks to ensure stability during loading. In this way, the wagon is converted into a solid bridge between the platforms for a length of over 23m; interlocks ensure that all the props are retracted before the shuttle departs.

With the bridging plates lowered, a lorry can then use the full width of both platforms, plus that of the wagon itself, to align its trailer or semitrailer with the first carrier wagon. When the shuttle is ready to depart the plates are raised to come within the loading gauge. This operation is checked by the train captain (chef de train) from his cabin in the club car.

During remarshalling in the terminals, a half-shuttle sometimes has to be propelled for some distance with the loading wagon leading. With the locomotive some 200m away, this would be difficult even at slow speeds and so a set of shunting controls is provided in the cabin at the end of the wagon.

BOGIES

The bogies used on the loading wagons are the same as those on the carrier wagons. Because the loading wagons always operate empty, their bogies are

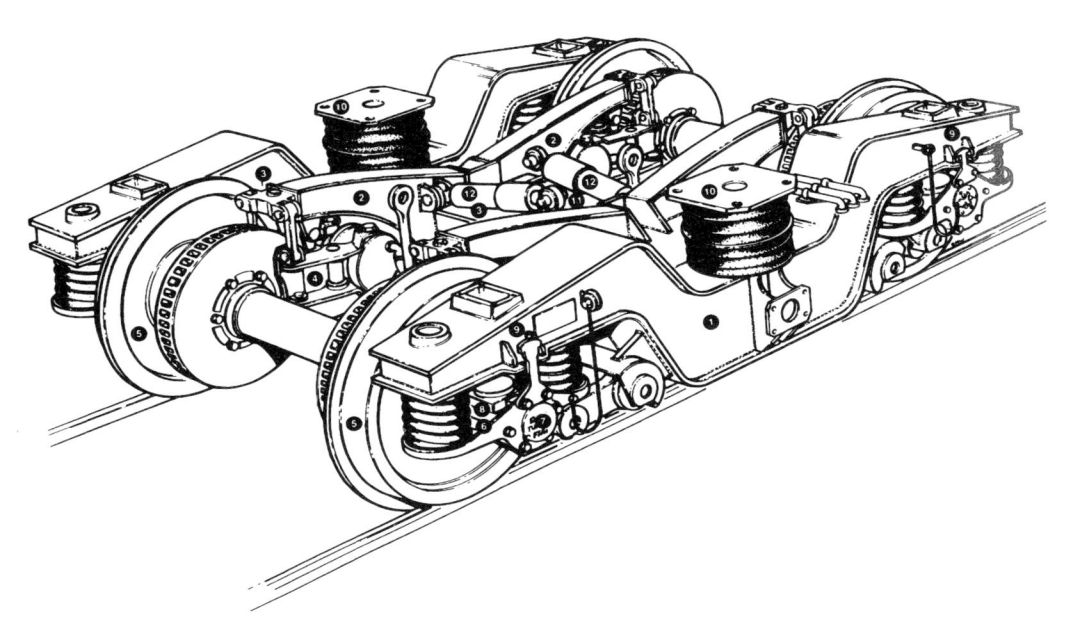

Figure 3. Freight shuttle: bogie (Source: Eurotunnel)

not fitted with load-sensors to control the braking forces. The heavy axle-loading and different suspension characteristics of the freight shuttle wagons led the builder, Fiat, to propose a lighter and cheaper bogie than that used in the passenger-vehicle shuttle wagons, although its design is similar.

As shown in Figure 3, the main frame consists of an H-shaped welded fabrication (1), with four shorter arms (2) attached to the cross-member (3). These support the brake gear (4), which is hung from their ends. The two wheelsets (5) are guided and attached to the main fabrication by short radial arms which link the roller-bearing axleboxes for all longitudinal forces. To support the greater weight, twin helical springs (6) are located vertically between the bogie frame and the roller-bearing support, which forms the primary suspension on each side of the axleboxes (7). The outer springs have hydraulic dampers mounted inside them (8). The rail-shaped pieces of steel (9) attached to the top of each axlebox normally slide loosely between guides on the main frame, and form stops to limit the vertical movement of the axle and permit the wagon and bogie to be lifted without losing the axles.

The secondary suspension is provided by a pair of hard rubber springs (10) forming the side links between the bogie frame and the vehicle. They are limited to a 32mm vertical extension. Bolted at their top to the wagon underframe, they enable the bogie to rotate about the drawgear pivot (11, omitted for clarity) on curves, and also provide some of the restoring forces to keep it parallel with the vehicle's under-frame on straight stretches of the track. They also act as noise dampers. Hydraulic dampers (12) horizontally mounted between the bogie and the underframe prevent the bogie hunting at higher speeds.

The rubber springs and the central draw-gear transmit the traction and braking forces between the bogie and the vehicle body. The drawgear is rigidly fixed to the underframe, and fits into the central slot in the bogie cross-member. The pivot deals only with the rotational movements, since the other forces between the bogie and wagon body are handled by different parts of the equipment.

As on other shuttle bogies, the brakes are air-operated electrically. The air cylinders move the ends of the callipers outwards, forcing the four brake pads into contact with both sides of the larger-diameter discs rigidly fixed to the axles. As is usual in railway practice, the discs are hollow, with integrally cast spacers between the two faces. At speed these act as fans, providing a flow of cooling air. Because the weight of any freight wagon varies widely (between 36 and 80 tonnes) depending on whether or not it is loaded, load-sensors and microprocessor controls regulate the braking forces, and antiwheelslide electronic speed control devices are fitted.

CLUB CARS

Lorry drivers and their passengers travel in their own club car, or amenity coach, marshalled just behind the front locomotive. An upmarket design was selected, based on the first-class passenger coaches used in the Italian State Railways' ETR 500 electric trainsets, designed to reach 270km/h on the 'Direttissima' high-speed lines. However, because such speed will not be attained in the Tunnel it has been possible to use a simpler bogie. The bogie chosen has been used extensively on other continental mainline rolling stock operating at speeds of up to 160km/h, and is basically the same design as that used on the freight wagons, but with softer springing.

A number of other changes have also been made to the basic vehicle to adapt it to its new role. Inside a section has been closed off to provide space for the train captain and the control equipment, while another area has been provided to accommodate a catering trolley. Because the club car is built to standard UIC loading gauge, it is much narrower than the other shuttle rolling stock, although the same width as the locomotives. The resulting gap between the car and the terminal platform (and the tunnel walkways if the train has to be evacuated) is bridged by a retractable flap, 670mm wide, that slides out below the steps provided at each doorway; a handrail also swings down from the end of the vehicle to provide assistance.

Each club car has seats for 52 passengers, arranged in pairs on one side of the offset aisle and in groups of four on the other. Tables with folding flaps are provided for each bay of seats. The coach is air conditioned; window blinds are supplied for passengers who find the contrast too great as they emerge from the Tunnel. Toilets are provided and a public address system is installed.

One feature of the Italian ETR 500 trains that has been retained in the club car is the continuous dark band of glazing along its length, although the actual windows are conventionally situated by each bay of seats. This is now a common design feature on railway stock; a similar effect was achieved by painting the upper panels of BR's InterCity stock dark grey to avoid the windows breaking up the lines of the coaches. All the vehicles in the train thus blend visually into a whole.

FIRE PREVENTION

In theory, a fire could break out in a lorry travelling on the shuttle, just as it could in a car on one of the passenger-vehicle shuttles. On the latter, passengers or train crew would raise the alarm; in addition the closed shuttle wagons are fitted with automatic fire detection sensors. The semi-open construction of the freight wagons means that effective fire detection equipment cannot be fitted. Instead detectors have been installed on the loading wagons. These are supplemented by additional detectors placed at intervals along the Tunnel, behind closed doors in the cross-passages where they can easily be maintained by staff working from the service tunnel. Small fans continuously suck a stream of air from the tunnels into each detector where it is tested for fumes and smoke.

Aerodynamics, ventilation and cooling

DAVID HENSON

The design of the Channel Tunnel posed unusual problems in terms of aerodynamics, ventilation and cooling. The tunnels are long; it takes a sound wave 2.5 min to travel from one end to the other. The shuttle trains are large and these, together with the through trains, travel at high speeds and at close headways. It will be one of the world's busiest railways with trains at headways down to 3 min, possibly lower in the future.

The three tunnels together contain nearly 6000 tonnes of air, which has to be refreshed and cooled to provide comfortable and safe conditions for the occupants. Travelling at high speed, the trains create large piston-effect pressure changes which could affect the comfort of passengers. They could also affect train ventilation systems, the structure of the trains, the equipment in the tunnels and especially the doors between tunnels, and the ventilation fans. Pressure changes from the train movement interact with the supply of ventilation air. The largest design pressure differences under consideration are 0.3bar, enough to lift a slab of concrete 1.5m thick.

The more aerodynamic drag on the trains, the more power from the locomotives is needed to overcome it. The greater the power need, the heavier the catenary needs to be, the greater the load on the traction supply system and the higher the cost of electrical power. Piston relief ducts are provided which greatly reduce the train drag. Energy is dissipated as heat which warms the tunnels and has to be removed by some means, and so affects the cost of cooling. This tunnel is the first main-line railway to have a mechanical cooling system.

The aerodynamic effects and the performance of the ventilation in incident and emergency situations are also very important, from temporary delays of trains in the tunnels to serious incidents with possible train evacuation and transfer of passengers to the service tunnel.

The tunnel systems require regular maintenance. A section of one tunnel may be closed for maintenance, usually overnight, and conditions for maintenance workers must be acceptable.

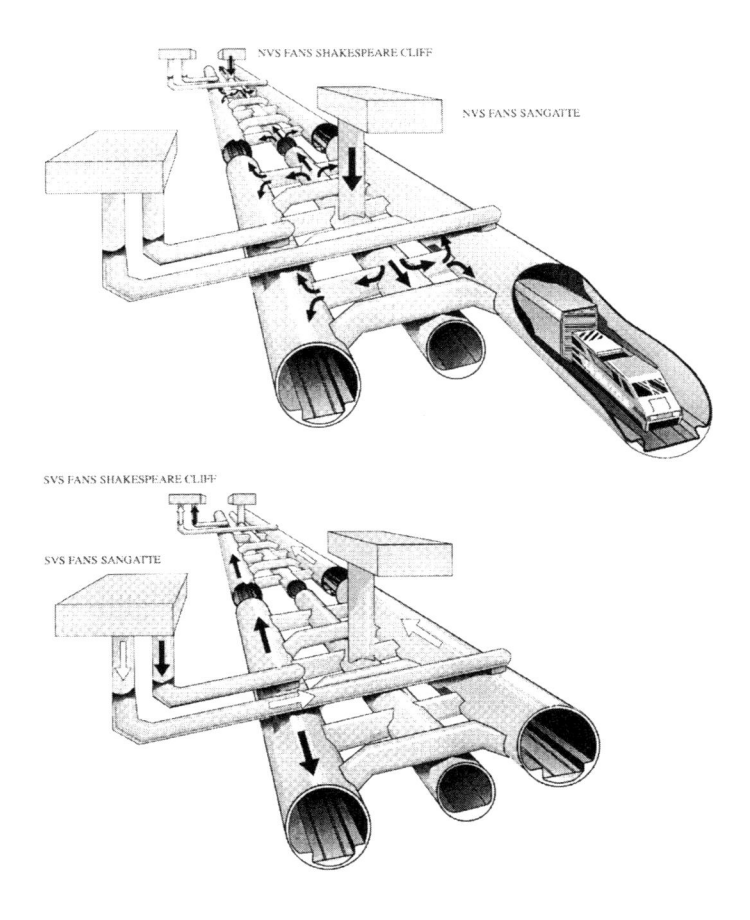

Figure 1. Top: normal ventilation system (NVS). Below: supplementary ventilation system

VENTILATION

The Channel Tunnel has two ventilation systems (Figure 1). The normal ventilation system runs continuously and ventilates the tunnels during normal train operation. The second is the supplementary ventilation system which is used in the event of an incident of any type that results in occupied trains coming to rest in the tunnels.

Normal ventilation

For design purposes it was estimated that the absolute maximum number of people that could ever be within the Tunnel at any one time is 20,000. At 26m^3/h per person this gives an air requirement of 144m^3/s. All trains using the tunnels have electric traction so there is no engine exhaust pollution in the tunnels; the service tunnel vehicles have diesel engines designed to give very low emissions.

The service tunnel is used as an air supply duct for the normal ventilation. This 4.8m-diameter service tunnel lies between the two 7.6m-diameter rail tunnels and is connected to them by cross-passages at 375m nominal spacing. The cross-passages are normally closed by doors (Figure 2) and air handling units installed above the doors control the flow of air from the service tunnel to the rail tunnels. The units are provided with nonreturn dampers to prevent air from the rail tunnels entering the service tunnel. The units can also be closed from the control centre. They are installed in every third cross-passage over most of the tunnel length, but in all of the cross-passages near to the centre of the tunnel.

Air locks, in the form of a pair of doors 55m apart, close the ends of the service tunnel to prevent the air escaping through the portals while allowing access by the service tunnel vehicles.

The piston (or pressure) relief ducts (Figure 3) provide a separate set of cross-connections between the rail tunnels at 250m spacing. These pass over the service tunnel and are not connected to it. The piston relief ducts are 2m in diameter and are fitted with butterfly dampers that are normally open but can be closed remotely from the control centre.

Figure 2. Cross-passage door. (Source: TML)

Figure 3. Piston relief duct (left) in rail tunnel; also shown are cross-passage towards right, walkways, cabling, pipework and lighting. (Source: QA Photos)

Air is supplied to the service tunnel from ventilation buildings at Shakespeare Cliff and Sangatte. At the English side the fan buildings are connected to the service tunnel by a shaft. There are two fans, each rated at 88m^3/s at a pressure of 2.9kPa. Each of the fans can do the full duty so that there is 100% standby capacity. This enables the ventilation to be maintained if one fan fails and also routine maintenance can be carried out on one fan whilst the other continues to operate.

On the French side, ductwork was installed in the very large (55m diameter) vertical shaft built at Sangatte to connect the fan building at the surface to the service tunnel. Here the fans are rated at 73m^3/s at 2.5kPa. They are not the same on each side as the tunnel lengths differ between the ventilation station at the coast and the portals.

Supplementary ventilation

A separate ventilation system operates through separate air shafts at each coast which supply or exhaust air directly to or from the rail tunnels. Thus, by operating from one side in supply and the other in exhaust a longitudinal air flow can be provided along either or both of the rail tunnels in either direction. The fan duties on the English side for each fan are 260m^3/s at 1.8kPa and on the French side 300m^3/s at 1.5kPa. Again, each fan can do the duty required but the fans can also be run in parallel to augment the air flow. These fans are fitted with variable pitch blades so that the duty cycle can be varied and are fully reversible. The fans are fitted with an antistall system to minimise the effects of pressure changes caused by train movements in the tunnels. All of the fan control, together with the closure of the dampers in the piston relief ducts and the air handling units in the cross-passages, is effected from the control centres at Folkestone and Calais.

If trains come to rest in the tunnels for any reason the train air conditioning continues to heat the air in the tunnel around the train. The normal ventilation rate is too low to provide a significant air flow over the train so the temperature would rise to unacceptable levels. The supplementary ventilation provides a longitudinal airflow to maintain acceptable conditions within the train for as long as necessary. Similarly if, for example, a traction supply failure caused trains to come to rest with the train air conditioning not operating, the supplementary ventilation would provide an air flow over the trains to provide local cooling.

In the event of an incident that causes trains to stop in the tunnel due to fire, the supplementary ventilation assists in the safe evacuation of passengers. The fans would be operated to provide an air flow in the tunnels so that the smoke from a fire would be moved in one direction along the tunnels leaving a smoke-free evacuation route for the passengers in the other direction. A velocity of 2.5m/s can be generated in the tunnel, which is sufficient to prevent the movement of smoke in the train annulus from moving upstream against the flow. This was verified using computational fluid dynamics techniques which showed that the smoke from the fire did not advance against the airflow in the annulus. The piston relief duct dampers would be closed to prevent smoke entering the other rail tunnel. Passengers would then evacuate to the service tunnel through the doors in the cross-passages. The normal ventilation system also assists in this situation. The supplying of air to the service tunnel causes the pressure within it to be greater than that in the rail tunnels and there is thus a flow of fresh air through the open doors. This prevents smoke from entering the service tunnel and assists passenger evacuation as people tend to move towards the fresh air flow.

The flow of air through the cross-passage doorways must be controlled to quite small tolerances in this situation. The minimum velocity required is approximately 2.5m/s. This is needed to provide a flow which can be felt by approaching evacuees and also to reduce the possibility of smoke entry. The velocity must also be kept below about 12m/s as above this level some people would begin to have difficulty in walking against it. Some degree of control of the normal and emergency ventilation systems is therefore required depending on where in the tunnels the incident train is located. Different fan settings are used if the train is near to the centre of the undersea section or if it is near to the coasts. Flow is further complicated by the movement of trains. Following an incident it would be necessary to bring out the trains not directly involved. Those ahead of the incident train would continue out of the tunnel while those behind would stop and reverse at low speed. Trains in the unaffected tunnel would continue out. An evacuation train will need to enter the tunnels to transport the passengers waiting in a stranded train or in the service tunnel.

The aerodynamic drag power of the trains is considerably greater than the power of the fans, so that while trains are moving they dominate the air flow in the tunnels, nullifying much of the effect of the fans. The air near the incident train would move erratically, reversing when a train passed in the other tunnel, so moving smoke over the evacuating passengers. The flow through the open cross-passage doors would also fluctuate, exceeding the maximum value and reversing in direction to cause smoke entry to the service tunnel. This situation is clearly unacceptable and it is therefore necessary to exercise precise control of the evacuating trains, doing so in a systematic way and at much reduced speed, down to 10km/h in some cases. During the design and construction stages a very large number of computer simulations were carried out to show the air flows generated by the ventilation fans and by the movements of trains and these assisted in the preparation of the operating procedures that will be used in these situations.

AERODYNAMICS

In its basic form the movement of a train through a tunnel causes the pressure ahead of the train to rise and the pressure behind to fall. This creates an air flow over the length of the train from front to back which, combined with the direct resistance of the pressure gradient, increases the aerodynamic drag of the train. In short tunnels, or tunnels with air shafts to the surface, the air in the tunnels accelerates to approach the speed of the train and so the overall effect on the train is relatively small. In addition, with short tunnels any slowing of the train due to the increased drag is not significant to the overall journey time. In the Channel Tunnel, at 50km in length, the pressure over the length of a single passenger-vehicle shuttle travelling at 160km/h would build to 20kPa with an aerodynamic drag of 35MW. With several trains in the tunnel at once the effect is lessened as the trains tend to assist each other in accelerating the tunnel air and both the pressure gradient and the aerodynamic drag are reduced.

The piston relief ducts act to reduce the pressure difference over the length of the trains. Air from the high-pressure region in front of the train flows through the piston relief ducts to the other rail tunnel. From there the air flows in the opposite direction past the train and returns through the piston relief ducts to the low-pressure region behind the train. These ducts reduce the pressure difference over the length of the shuttle to 6kPa and reduce the relative air velocity alongside it, which in turn reduces the train drag. The flow regime is such that, as two trains pass, the circulating flow pattern of each is merged and reinforced so there is further reduction in the pressure drop over the train and consequently in the drag. The effectiveness of the ducts is greatest with trains at longer headways. At very low headways, in a tunnel without ducts, the trains assist each other through the tunnel and similar reductions of drag occur.

Optimisation of piston relief ducts

Repeated aerodynamic calculations have shown that the reduction of train drag effected by the piston relief ducts is closely proportional to the cross-sectional area of duct per kilometre of tunnel. It has therefore been possible to optimise the number and size of the piston relief ducts. The cost of construction is roughly proportional to the total cross-section of piston relief duct provided (with variation depending on such factors as the number of lining rings affected) and this cost can be converted to an annual cost over a number of years. The aerodynamic power reduces with increased piston relief duct area. The reduction of drag reduces the cost of locomotives, catenary, cooling system and the electrical power for traction and cooling. This can also be converted to an annual cost. Summing the curves gives a curve with a minimum value which represents the optimum at approximately 12 m²/km of tunnel. Other factors, including construction, pressure pulses affecting passengers etc., determined the chosen arrangement of ducts of approximately 2m diameter at 250m spacing.

Under this arrangement of piston relief ducts Figure 4 demonstrates how the power required by the locomotives varies with speed. The diagram shows

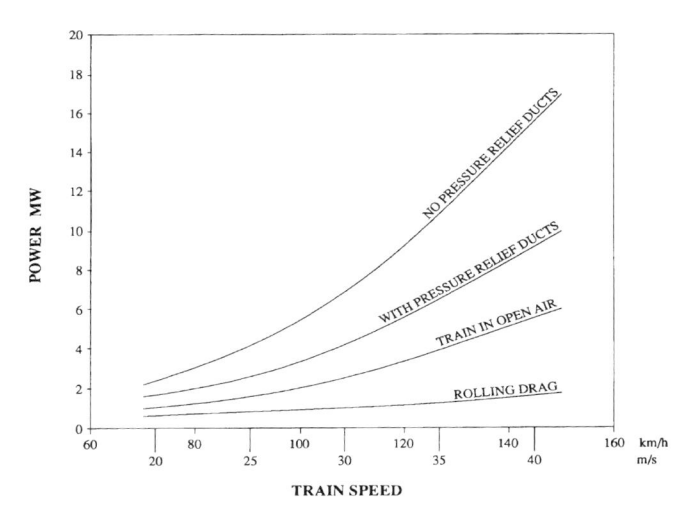

Figure 4. Train power/speed

the power needed with and without the piston (or pressure) relief ducts, together with the power in the open air and the rolling resistance (i.e. the total train drag without aerodynamic resistance).

Lateral forces

One disadvantage of the piston relief ducts is the sideways or lateral force exerted on the trains by the strong air flow through the piston relief ducts away from the train near to the front and a lesser flow towards the train at the rear. The top corners of the shuttle wagons, which are generally of a square cross-section in a circular tunnel, restrict the flow of air from one side of the train to the other, the large flows through the piston relief ducts generate a local imbalance in the pressure on each side of the train and a sideways force is generated. The force affects ride comfort, and in extreme cases could cause excessive sideways movement of the wagon to violate the structure gauge.

The jet of air that blows onto the side of the train near to the rear also affects soft-bodied freight wagons operated by BR and SNCF. These wagons have a light frame which supports a tarpaulin cover. Tests were carried out which showed that the cover could be damaged by the air jet. Eurotunnel therefore took the precaution of adding flow restrictors to the piston relief ducts to reduce the maximum flow rate through them. The restrictors will be removed at a later date if the problems associated with these wagons can be avoided. In the meantime the aerodynamic drag of the trains will be increased, leading to higher costs of power for traction and tunnel cooling, estimated at about 4%.

Organ pipe effect

Questions were raised concerning the so-called organ pipe effect. It was suggested that pressure wave oscillations could be set up in the sections of the Tunnel between piston relief ducts, causing some disturbance to the tunnel occupants. This aspect was studied by acoustic specialists who determined that the geometry and necessary conditions were not in fact right for this effect to occur.

Pressures affecting passengers

The effect of the changing air velocity in the rail tunnels as the trains pass the piston relief ducts generates pressure fluctuations. Extensive tests were carried out by British Rail Research at Derby to determine whether these and other pressure changes caused by interaction of trains would be noticeable or disturbing to passengers and train crews.

A number of volunteers sat in a cabin in which the pressure was varied to simulate a typical journey through the Tunnel. The pressure history used was the calculated annulus pressure for a shuttle train operating with other shuttles and through trains. The subjects then recorded the acceptability of the pressure changes during and after the journey.

Before these tests were carried out, it had been proposed to leave open the supplementary ventilation shafts at the coasts to supplement the natural ventilation. As a result of the tests it was decided to close these shafts because the pressure pulses generated as the train passed the shafts were the most noticeable. Subsequent tests with the shafts closed gave acceptable results.

The air pressure design criteria were derived from these tests and similar tests in other tunnels. The pressure change for isolated pressure changes is 3kPa acting over less than about three seconds. Repeated pulses depend on the frequency, reducing to 0.45kPa for the pulses generated as the train passes each piston relief duct.

Single-line working

During periods of maintenance one of the rail tunnels may be closed and alternate two-way working will operate in the other. Crossovers are provided between the rail tunnels so that only approximately one third of one tunnel needs to be closed at any time. Large longitudinal doors were provided in each crossover chamber so that the rail tunnels are separated aerodynamically during normal operation (see Chapter 22, Figure 1). These doors are very large at over 300m^2. To maintain continuity the service tunnel diverts past the crossovers.

If the piston relief ducts were left open the air velocity in the closed tunnel would be too high for maintenance to be carried out. The air velocity would reach approximately half of the train speed, or 20m/s, which is too great for maintenance personnel. It was therefore necessary to install dampers on each piston relief duct to close them during maintenance. This of course removes the pressure relief effect of the ducts and it is necessary for trains to run at reduced speed to reduce drag, pressure and velocity effects.

Effect on ventilation air supply

The pressure in the tunnels varies considerably as the trains move along the tunnels, which influences the distribution of the normal ventilation air entering the rail tunnels from the service tunnel. Nonreturn dampers are incorporated into the air handling units in the cross-passages. When the air pressure in the rail tunnels is higher than that of the service tunnel the dampers close. When the pressure is lower the dampers open and air is drawn into the rail tunnels. This means that at any one moment the air flow from the service tunnel into the rail tunnels is very uneven. However, over a period of a few minutes the air flow to the tunnels balances out and the overall effect is an even distribution of air along the full length.

AERODYNAMIC CALCULATIONS

The design calculations for the aerodynamics and cooling of the tunnels were done using computer programs at Mott MacDonald. The tunnel aerodynamic program uses compressible, unsteady flow theory solved by the method of characteristics and was used extensively during the design process.

Basic data for the calculations fell into two main categories. The first category included items which were known or fixed by design, for example tunnel diameter, train speed etc. (Table 1). The second category contained items which were unknown and required estimation or determination by measurement and testing. These included tunnel and train friction factors, loss coefficients for each end of the trains, losses at the piston relief duct junctions, amounts of tunnel seepage water and others. Extensive tests were carried out at both model and full scales to determine these empirical values.

The rail tunnel friction factor was of importance to the aerodynamic effects. The tunnels are lined over the majority of the length with concrete segments which are internally smooth. Some sections, particularly at cross-passage and piston relief duct connections, have cast-iron linings which are internally ribbed. Services, including power and control cables, pipework, catenary

Table 1. Basic data used in aerodynamic calculations*

Tunnels	
Tunnel length	50,193.0m
Cross-sectional area	42.2m^2
Perimeter length	26.8m
Piston relief duct diameter	2.0m
Piston relief duct spacing	250.0m (nominal)
Shuttle trains	
Length	790.0m
Cross-sectional area	21.2m^2
Perimeter length	18.0m
Speed	varying to 160.0km/h
Through trains	
Length	427.0m
Cross-sectional area	8.6m^2
Perimeter length	11.4m
Speed	varying to 160.0km/h

*May not exactly represent the 'as constructed' values

system etc., are distributed between the tunnels and, together with the track, contribute significantly to the tunnel friction.

Given relevant data for the system geometry, design parameters, fan and train details etc., the program calculated the air pressure and velocity in all parts of the tunnel network as they varied with time. Other output provided the train drag and power needed to overcome it, and results were produced both numerically and graphically.

Information on the air flows and heat gains from the trains formed the basis of the thermodynamic calculation, together with information on the linings, surrounding ground, cooling system and water infiltration. The program calculated the temperature in all sections of the tunnel as this varied with time over periods of days, months and years.

The calculations for the Channel Tunnel developed into an iterative process as the design proceeded. It would not be practical to calculate every possible option of tunnel geometry and train arrangement, so the principle of a 'reference design' was used. This employed a basic set of data for the tunnels, trains and fans for which results were calculated. Modifications were made to the parameters in turn to show the effect of the changes for comparison purposes. At convenient times in the overall system design process the reference design was altered to reflect more closely the current design of the system.

The importance of using compressible flow theory was demonstrated in the velocity profile where the velocity along the length of the annulus varies by several metres per second due to the changing pressure. This causes a higher air speed in the tunnel ahead of the train compared with the air velocity behind.

Commissioning tests carried out prior to the Tunnel opening included the measurement of air pressure and velocity at various points in the tunnels as trains passed. The results of the measurements were compared with the earlier calculations and the agreement was found to be very good. Any variation was due to the inability to define precisely the train position with time so that the calculation would match the measurement. In addition, the calculation assumed that the tunnels were completely airtight whereas in practice it was not possible to achieve this. There is some leakage of air between the three tunnels, partly at the crossovers, cross-passage doors and dampers but also at equipment rooms where these link to both the service tunnel and the rail tunnels. The amount of leakage was reduced during the commissioning period.

Calculations for normal train operation

A large number of calculations were done for this project. Typical results are shown for a shuttle train running under simulated 'normal' conditions among other shuttle and through trains. Train speeds were varied to simulate actual running speeds at different parts of the Tunnel. Figure 5 shows the pressure in the tunnels at the nose, each end of the annulus, and behind the train for one shuttle during a complete transit. The pressure difference over the length of the train varies up to approximately 6kPa but the actual pressure varies over a larger range depending on the position of the other trains in the same tunnel and in the other rail tunnel. There is generally a drop in pressure over the train when trains pass. The inset on the graph shows a section of the results in more detail. The smaller but frequent pressure fluctuations occur as the train passes the piston relief ducts.

Figure 6 shows the air velocity in the tunnel and annulus for the same calculation. In this example, the air velocity in the tunnel ahead of the train remains relatively constant at about 18m/s, which is just over half of the train speed, while the velocity in the annulus generally remains at less than −5m/s relative to the tunnel.

The instantaneous power required to overcome aerodynamic drag is approximately proportional to the cube of the train speed and is therefore very sensitive to the speed (Figure 7). The maximum train speeds would be attained on the down gradients while towards the end of the transit, on the up gradients, the speeds are less and the aerodynamic drag is significantly reduced.

Figure 8 shows the train paths of all the trains in this calculation. The X axis represents the length of the tunnel and the Y axis shows time. The pairs of lines represent the two ends of each of the trains as they move through the tunnel. The longer of the trains are the shuttles whilst the shorter trains represent the through trains. In this example a sequence of two shuttles followed by a through train and then a free path has been used. Results are shown for the train the enters the UK portal (at the left of the diagram) at the axis time of 3 min.

An example of the instantaneous pressure along the length of the eastbound tunnel at a fixed time is shown in Figure 9, corresponding to a time of 7.5 min in Figure 5. The inset shows the position of the trains, the results being shown for the upper tunnel where trains are moving from left to right. The large pressure steps represent the shuttle trains and the smaller steps are caused by the through trains which have a considerably smaller cross-sectional area.

Exceptional train arrangements

There is a large number of possible train arrangements and speeds in normal and single-line working and much of the design effort involved the determination of the worst-case design. Having decided on a particular train

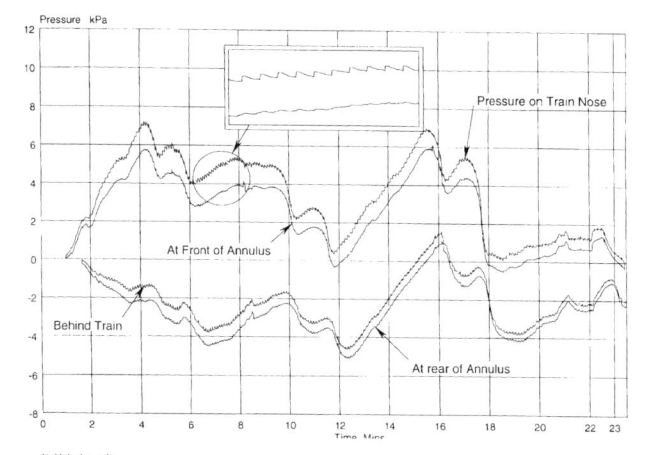

Figure 5. Air pressure near typical shuttle

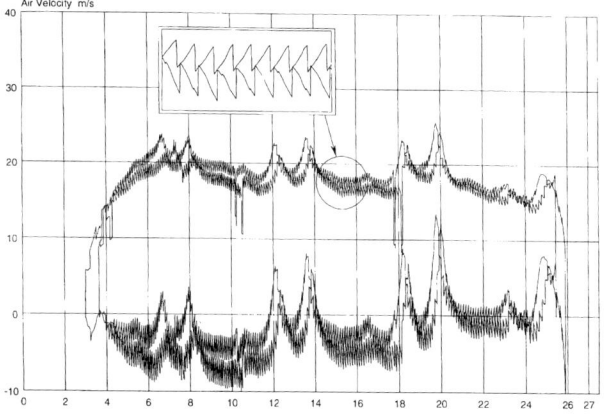

Figure 6. Air velocity near typical shuttle. Upper lines represent air velocity in the tunnel ahead of the train; lower lines the velocity in the annulus

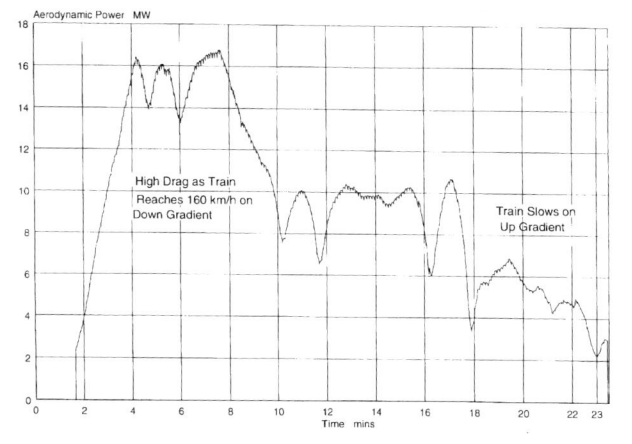

Figure 7. Power required to overcome aerodynamic drag

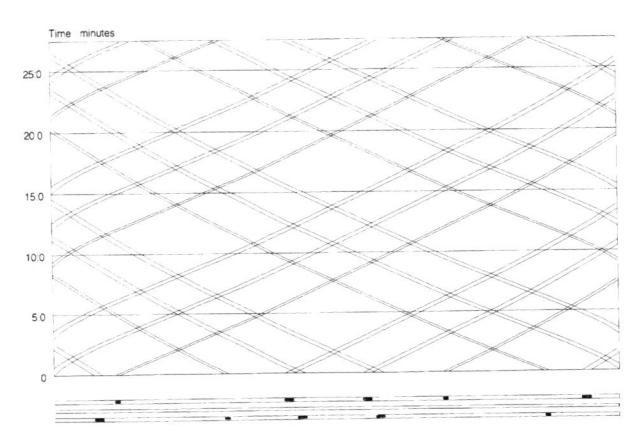

Figure 8. Train path diagram. The X axis represents the length of the tunnel and the Y axis shows time. The pairs of lines represent the two ends of each of the trains as they move through the tunnel

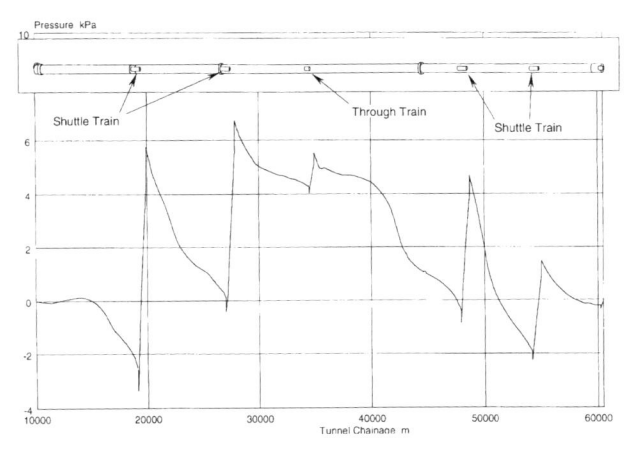

Figure 9. Typical pressure profile in tunnel

arrangement it was then necessary to decide if it could occur in practice and whether it was realistic to design for it, or whether restrictions should be placed on train movements to avoid the particular event. The points of interest included the maximum velocity that could occur in a piston relief duct and the maximum pressure differences between the tunnels which would affect the cross-passage doors and the piston relief duct dampers.

One example occurs when a flight of shuttles approaches the centre of the Tunnel from each direction at minimum headways. The trains act to compress the air in the centre of the Tunnel to a pressure of 15kPa. This would not affect the passengers as the buildup of pressure would be relatively slow. Another case occurs if one piston relief duct damper fails to close when all the others are closed. This could lead to very high velocities in the open duct.

The design criterion for the maximum pressure difference that could occur was set at 30kPa and the cross-passage doors, the crossover doors and the piston relief duct dampers were designed to withstand this pressure.

Rolling stock

The Tunnel aerodynamics affect the design of the rolling stock in many respects. The high train drag influences the power required by the locomotives, and consequently the power supply system. The power required by the train in the Tunnel is much greater than that required in the open air. The drag is, however, greatly reduced by the piston relief ducts, although these introduce their own problems, as outlined above.

It is generally not practicable to construct airtight trains, particularly large shuttle wagons, and so the leakage causes the pressure inside the train to balance at the mean of the pressure in the annulus outside. As the shuttle train moves along the tunnel there is a considerable difference in pressure between the two ends of the train, typically 6kPa. If the shuttle train were nearly airtight and all of the internal doors between the wagons were open the internal pressure would balance at the mean of the pressure in the annulus. The pressure difference between the inside and outside of the wagons near to the ends of the train would then be of the order of 3kPa, and would require substantial construction to avoid collapse of the end wagons. It would also be very difficult to control the ventilation within the train. With typical amounts of leakage that could be expected a large quantity of air would enter the train in the front part and leave at the rear so generating internal winds near to the centre of the train.

This effect was demonstrated in a passenger train during tests carried out by Eurotunnel in the Simplon Tunnel between Switzerland and Italy. The internal doors of the train were held open and a velocity of 12m/s was measured near the centre of the train. When the door was closed it was not possible for one person to reopen it. If internal doors between each wagon were closed the internal pressure of each wagon would balance at the mean of the external pressure alongside it. The structural loading on each wagon is then considerably reduced. However, the rapid pressure changes within the tunnels can cause significant loading on the wagons and fatigue considerations play an important part in the wagons' structural design.

The pressure difference between each wagon is now quite small but there is a pressure loading on the doors between the wagons. This could be large enough to make opening a door difficult. Care is needed to ensure that a number of consecutive doors are not opened at once. If a significant length of the train were open, winds could be generated within the train and internal doors could become difficult to open or close. It is therefore necessary for trains within the Tunnel to have the majority of interconnecting doors closed during transit.

Fixed equipment

The pressures and air velocity generated by the movements of the trains were taken into account in the design of all equipment in the tunnels. This includes the services in the tunnels, including cables and cable trays, cooling pipes etc., and such items as the drainage channel covers. The pressure differences affect the design of the crossover doors and cross-passage doors (normally all closed) and a design pressure of 30kPa has been used. The piston relief duct valves are normally open.

COOLING

It is not usual for railway tunnels to have mechanical cooling equipment. A number of rapid transit systems have refrigeration for cooling at stations, including Hong Kong and Singapore, but the Channel Tunnel is the first mainline railway tunnel to have special cooling equipment.

The design limit for temperature in the rail tunnels was set at 30°C. It could be argued that, as most passengers travelling in the Tunnel are in air-conditioned vehicles, the design limit could be higher. The limit was set to allow for passengers to be able to detrain if necessary in emergency conditions and to make maintenance conditions acceptable.

The trains generate a large amount of heat in the tunnels. The traction equipment is rather less than 100% efficient and this causes the equipment to get warm. Similarly the power needed to overcome the mechanical resistance within the train and the aerodynamic drag is lost as heat, together with the energy from auxiliary systems, including air conditioning and lighting. In general terms all of the electrical power supplied to the trains is ultimately dissipated as heat. There are additional gains from electrical equipment in the tunnels and from the catenary.

In other tunnels the heat deposited by the trains is generally removed by air exchange. The piston effect of each train causes air movement within the tunnel with consequent air exchange at the portals and air shafts. In the Channel Tunnel there is very little air exchange with the outside. Much of the air circulates between the rail tunnels through the piston relief ducts. The amount of air leaving at the portals is quite small and relatively little heat is removed by this means from the system. The piston effect of the trains circulates air through the portals and the first few piston relief ducts and generates some local cooling near to the ends of the tunnels.

During the early stages of operation of the Tunnel a large amount of heat will be absorbed by the tunnel linings and surrounding ground. However, as these reach equilibrium temperature the amount of heat removed from the tunnels will be only a small percentage of the heat generated.

Water is also a good vehicle for the removal of heat. Water infiltrating into the Tunnel would warm up and so remove heat when it is pumped out of the Tunnel. Before construction began the amount of infiltration water was unknown, and it has subsequently been shown that much of the Tunnel is dry. Water entering the Tunnel would be channelled to the sumps and so the amount of heat absorbed was unknown. For these reasons water was not included in the calculations as a reliable mechanism for heat removal.

Heat can also be absorbed by the structure of the trains. Trains entering the Tunnel during cool ambient conditions warm up in the Tunnel and so remove heat. However, during very warm weather this process would be reversed and cause a heat gain to the tunnels. This mechanism cannot therefore be used as a reliable means of heat removal.

Taking these factors into account it was shown that the heat gains to the Tunnel were considerably greater than the natural heat removal mechanisms available and it was necessary to increase the heat removed by other means.

A large number of methods were compared and evaluated. The cheapest and best method for many railway systems is to provide air shafts to the surface at regular intervals. These increase the air exchange and so remove the generated heat, but they also serve a similar function to the piston relief ducts, removing the need for such ducts. However, the construction of ventilation shafts within the Channel was considered impractical, partly because of the high cost but also because of the dangers of impact by shipping. A very large amount of protection would be needed around an air shaft to prevent damage to the shaft which might cause flooding of the tunnels. There would also be additional hazard to the shipping, the English Channel being one of the busiest shipping routes in the world.

Various methods were considered, some less practical than others. Spraying water into the Tunnel would not be effective. The evaporation of water reduces the sensible temperature at the expense of increasing the humidity, so the benefit to comfort is minimal. There is also not enough air exchange for the humid air to be removed, so the total amount of cooling is small. The addition of water could remove heat but creates problems of corrosion in the Tunnel equipment and in the linings.

Other less useful suggestions included having a tanker of ice on each train which would release ice into the Tunnel during the journey. Running the supplementary ventilation overnight was also proposed, but this would only remove 2–3% of the heat added during the day.

Figure 10. Refrigeration plant at Shakespeare Cliff. (Source: TML)

It was necessary, therefore, to provide a mechanical cooling system. The system installed has an initial capacity of 58MW, and uses chilled water circulating in pipes in the rail tunnels. Two large refrigeration plants were constructed, one at each coast (Figure 10). Chilled water, supplied at 3°C, is circulated in the pipes, arranged in the form of a hairpin, one in each of the underland section and one in each half of the undersea sections. In this way the water returns to the same plant. The pipes installed are of 400mm diameter in the undersea section and 300mm diameter in the underland sections. The pipes are installed in the two rail tunnels, on the side opposite to the walkway and cross-passages. The pipes are fitted with valves at regular intervals so that they can be closed in the event of pipe failure, to minimise the amount of water entering the drainage system.

It was shown that 600mm-diameter pipes will be needed when the ground around the Tunnel warms to the equilibrium temperature and when the traffic approaches the capacity of the tunnels over most of the day. The cooling load will then rise to 80MW. However, as these conditions are not anticipated for some years, possibly close to the design life of the pipes, it was decided to install the smaller pipes at the outset and replace them with the larger pipes, for which space has been provided, at a later time. The diameter required to contain the quantity of water needed is large enough for the surface area of the pipes to be sufficient for heat exchange without fan coil units or the assistance of any form of fins or vanes.

The outside air has little effect on the conditions in the Tunnel. This means that the temperature in the Tunnel and the cooling load is not directly related to the time of year so that the temperature within the tunnels will remain relatively constant. The cooling load will reduce in winter if there are fewer trains using the Tunnel. Also, the effect of the trains themselves entering with lower skin temperatures will provide some additional cooling.

No specific means are used to control humidity in the tunnels; it is probably not feasible to do so. Condensation onto the cooling pipes will generally keep the relative humidity below about 50–60%, and will also assist in the heat transfer to the pipe.

No specific minimum temperature was set for the tunnels. Precautions have been taken against freezing only very close to the portals and at the air supply points.

16

Fire detection and suppression

ERIC H WHITAKER

The one safety issue affecting the Channel Tunnel that has attracted more attention than any other is the risk of fire on a shuttle train, particularly a passenger-vehicle shuttle. In its own Safety Case published in 1994, Eurotunnel singled out a vehicle fire as the most significant risk to be addressed.

There is no historical evidence to suggest that the carriage of vehicles on trains is particularly hazardous. The vehicle shuttles through the Alpine tunnels have operated in complete safety for more than 30 years. However, Eurotunnel, the Intergovernmental Commission (IGC) and the Safety Authority gave this issue particular attention for three reasons:

- There was opposition, led by the ferry companies, to the principle that passengers should be allowed to remain with their cars on the passenger-vehicle shuttle (the 'nonsegregation concept'). They saw this as a competitive advantage enjoyed only be Eurotunnel.
- The fire statistics collated by the Home Office showed that fires, mostly in cars, had doubled in ten years
- The unusually long length of the Tunnel.

These last two reasons led fire professionals to add their concerns to those of the Tunnel's competitors, despite reassuring evidence to Parliament by both the Chief Inspector of Fire Services and the Chief Inspecting Officer of Railways.

Eurotunnel responded to these concerns by appointing a senior adviser (this author) with long fire service experience to assist them, and launched an extensive programme of research, testing and innovative design to meet its objective of producing the safest transport system in the world.

GOOD FIRE PREVENTION PRACTICE

There are ten general rules of good practice in fire prevention and design which, if adopted in various combinations, will ensure a satisfactory fire safety standard. All ten underpin the Eurotunnel system:

- using materials which would not burn or give off smoke
- subdividing vehicles into separate units so that any emergency would be self-contained
- providing a level of fire resistance so that a fire could be contained as long as necessary to safeguard human life
- having a system of detectors to identify fires, petroleum vapour or liquid petroleum gas (LPG)
- instituting methods of preventing or automatically extinguishing such fires once they reached a predetermined level.

This design would be backed by:

- an evacuation alarm system
- satisfactory means of evacuation for all scenarios
- clearly specified procedures and staff trained to carry them out
- the provision of staff procedures to ensure rules are observed and to respond to emergencies and assist evacuation, and finally
- a speedy response by local authority emergency services.

The basic concept of the Eurotunnel system effectively ruled out the possibility of separating passengers entirely from the risk of fire, for example by separating motorists from their cars.

VEHICLE FIRES

The starting point for Eurotunnel in planning and developing its safety design for the rolling stock was to assess the nature of fires in the vehicles it proposed to carry. It commissioned the UK Fire Research Station to produce three reports, an action which proved to be of considerable assistance in substantiating the nonsegregation concept.

The most important of the three reports analysed the UK Fire Statistics for vehicle fires and gave details about causes, and the proportion of each type of vehicle included in the 50,000 vehicle fires which occurred annually. The most important conclusion concerned the inability of the fire report format to provide some important facets of fire data required, so Eurotunnel asked Kent Fire Brigade to provide more comprehensive details of vehicle fires over a period of a year. This joint venture allowed much more detail to emerge, and the database was revisited many times during the following years. Although the analysis showed that cars in a stationary mode using the shuttle were ten times less likely to have a fire than when they were on the road, it did not provide data on the fire load of a car or the speed at which such fires develop. The next part of the research programme had to address this issue.

Three series of fire tests took place at the Mines Research Establishment in France over a period of years. A mock-up of a shuttle wagon was built and a programme of tests developed at which members of the Safety Authority, TML and the Maître d'Oeuvre were present. The first tests provided base data on the way cars burned both in open-ended shuttle wagons, and for comparison in the enclosed wagon. Information was obtained on the behaviour of detectors and fire suppression systems. In the second set of tests a single-deck mock-up was used to obtain data on coach fires, and this series of tests incorporated a simulation of the shuttle ventilation regime.

The test protocol also included an analysis of the environmental conditions in the test wagon during the fire test and afterwards. This was required for two reasons: first, to ascertain the types and levels of gases present as a direct result of the fire, and secondly to assess the way in which the preferred extinguishing agent, Halon 1301, reacted in the fire, a subject which was to occupy the minds of the Safety Authority for a considerable time. The third set of tests used a wagon form with detection and suppression equipment which was considered very close to the final specification (which is shown in Figure 1).

Figure 1. Attendant's control panel and fire extinguisher, passenger-vehicle shuttle. (Source: QA Photos)

The monitoring of these tests provided data which was subsequently used by the UK Fire Research Station to validate computational fluid dynamics computer models.

After each of these sets of tests a detailed analysis of the data was provided for TML, Eurotunnel, the Maître d'Oeuvre and the Safety Authority to consider, following which meetings were held to decide on those matters that had been resolved and those that still required additional work. When this costly series of tests ended there was a need to consider additional variations of fire scenarios. To simulate them, the UK Fire Research Station were eventually chosen to use their computer fluid dynamic model Jasmine to carry out this work. Work which incorporated scenarios that had not been used during the actual tests was undertaken, and this contributed to convincing the Safety Authority that the nonsegregated design was acceptable. At the end of this work it was possible to specify a fire incident and actually to see on a VDU how the fire and smoke developed, and to assess the time at which various detectors would operate.

Part-way through this process the Safety Authority considered that the information on the fire load and rate of fire development in modern cars was not sufficiently well documented, and they commissioned two such tests at the original Cardington airship hangars which were being used by the UK Fire Research Station. In the first test the fire burned rather too well and had to be aborted at its peak; otherwise the calorific testing equipment would not have been available for the second test. The test enabled the necessary data on fire loads of cars to be obtained and also highlighted the totally different manner in which the car fire developed in open conditions in contrast to an enclosed shuttle wagon.

The Safety Authority had given qualified approval for Eurotunnel to continue the concept of the nonsegregated design and the conditions they attached were the 'binding requirements'. Work continued over the next four years on modifying the design to incorporate these items including the provision of video cameras in each wagon, increasing the width of pass doors in the wagon shutters, making provisions for the disabled, special conditions for the carriage of coaches, and performance criteria for the detection and suppression systems.

SPECIAL VEHICLES

One of the decisions the Safety Authority made in connection with segregation was that they would not allow caravans, camper vans, vehicles with trailers or vans to be carried in the same tourist shuttles as passengers remaining with their vehicles. This group of vehicles became known as special vehicles. Eurotunnel realised that this category of cross-channel traffic had important revenue implications, and it decided to analyse the reasons for the Safety Authority's decision, to determine whether a convincing case could be made to alleviate their concerns.

The Safety Authority was primarily worried about the hazardous types of material that could be carried in vans and trailers, notably the cylinders of liquefied petroleum gas (LPG) that serve as a fuel in caravans and camper vans.

The first solution that was considered was a segregation of passengers and vehicles, i.e. by loading these special vehicles at the rear of the passenger-vehicle shuttle rakes and using a small bus to carry the passengers. However, this seriously affected the dwell times for shuttles at the terminals, was uneconomic in the loading of shuttle wagons, and was rejected as inappropriate. The second attempt to overcome the Safety Authority's objections revolved around modifications to procedures, which provided a better safety assessment as far as risks were concerned.

By carefully analysing fire statistics it was possible to identify the frequency of fires and explosions caused by LPG. The small number of explosions occurring in the UK in caravans and camper vans were examined again; many of these were the result of fires becoming uncontrolled and large enough to overheat the LPG cylinders. These circumstances would not occur in the shuttle wagons because of the fire detection and extinction system (FDE). The other group of explosions occurring as a direct result of the LPG gas itself being ignited were occasioned by operation of LPG cooking, heating and lighting systems. It was therefore decided that the prohibition of LPG cylinder valves being in the 'on' position during Tunnel transit should prevent such incidents, and Eurotunnel incorporated a rule that all caravans and camper vans would be inspected prior to loading.

In the case of light commercial vehicles, inspecting all of them for dangerous goods prior to deciding on whether they should be carried in the segregated or nonsegregated mode was thought to be too great a task. However, in examining the problem and surveying traffic at the ports it was clear that a greater number of vans and light commercial vehicles were being used as private vehicles by families. Others were being operated as commercial vehicles part of the time and for leisure and holidays on other occasions. As the private use of such transport posed no greater risk than the car equivalent, it was decided to propose that light commercial vehicles not carrying commercial goods should be considered as nonsegregated traffic.

Reports were produced comparing the risk of special vehicles with the risk associated with the already accepted classification of nonsegregated vehicles. With some qualifications, these were accepted by the Safety Authority which proposed to the IGC that the original decision should be reconsidered.

HGV ROLLING STOCK

The other area in which considerable research and testing took place in fire engineering was in the design and commissioning of the shuttle wagons carrying heavy goods vehicles (HGVs). The difficulty in adopting a completely enclosed design of wagon, similar to the passenger-vehicle rolling stock, was that the 44-tonne vehicles which the Channel Tunnel Concession prescribed made the total weight too great for existing types of bogies.

The adoption of a semi-open concept meant that many of the good fire prevention principles listed earlier could not apply. This would have the effect of nonisolation of a fire within a single wagon and a probability of spread from one HGV to the next. It would also allow smoke and combustion gases to enter the tunnel system, an aspect which was extremely unlikely with the enclosed passenger-vehicle wagon design. Both of these undesirable factors and many others were pointed out by the Safety Authority, Maître d'Oeuvre and safety engineers within Eurotunnel. However, the problem had to be solved and an outline strategy was proposed.

As life protection is the first priority in safety, the starting point was to consider the safety of the HGV drivers, passengers of following trains, and the emergency services which would have to deal with the consequences of such emergencies. It was decided to carry the drivers in a specially designed club car (or amenity coach) at the front of the HGV shuttles. The club car would have to give protection from fire until the whole train could be brought to the surface, or until such time that the drivers could be evacuated either by uncoupling the club car from the remainder of the shuttle and proceeding out of the Tunnel, or in more extreme cases by stopping the shuttle and evacuating into the service tunnel.

This concept had to address a number of issues. How could any fire on a HGV shuttle be detected? How fast would such a fire develop in a wind speed of that of the shuttle? How could trains following the affected HGV shuttle and those in the other rail tunnels be protected from the smoke?

The safety of trains in the other rail tunnel and those following the affected HGV shuttle was protected by the adoption of two important sets of procedures. The first set is known as the standard procedure for fires in trains in the tunnel and involves preventing further trains entering the Tunnel system, modifying speeds of trains, stopping those following the affected train, and completely separating each of the rail tunnels from one another by closing the dampers in the piston relief ducts. The second procedure incorporates a minimum headway between an HGV shuttle and the following train of 4km.

By producing a series of calculations for a range of emergency situations it was possible to show that the procedures would work to the satisfaction of the

Safety Authority. However, assumptions made about the rarity of such events and the rate of burning of HGVs had to be researched further. The probability of the event of a HGV fire while stationary with its engine off was determined by revisiting the Kent Fire Brigade vehicle fire survey. The information about the size and rate of development of HGV fires was a larger issue and this was resolved by further research.

In the late 1980s fire engineers throughout the world were aware of the absence of information and data on tunnel fires, and were experimenting in America and Europe in programmes completely unconnected with the Channel Tunnel project. The one in Europe was sufficiently far advanced for Eurotunnel to make an approach to see if it could be expanded to incorporate an HGV fire. Two engineers dealing with the HGV issue, Simon French from Eurotunnel and an ex-UK Fire Research Station engineer acting in a consultancy capacity, Bill Malhotra, managed to persuade the European Eureka project team to allow an additional HGV test at the end of the programme – providing the HGV could be delivered to the point in Norway above the Arctic Circle (Hammerfest) before the weather deteriorated in November. The test was made and useful data obtained about fire size and development factors which were previously a matter of conjecture. More importantly, valuable information was now available on the way in which the supplementary ventilation system was likely to control the flow of smoke for the safety of drivers in the HGV club car if, for any reason, they had to evacuate to the service tunnel. This information was also important to the Fire Service which may need to enter the affected rail tunnel to deal with fires on trains.

The Eureka project had another additional bonus, when the Health and Safety Executive testing station at Buxton was able to conduct scale model tests using some of the data to assess other types of HGV incidents. This helped to prove more comprehensively that the ventilation system was capable of controlling smoke flows. One of the most important findings in these tests indicated that scientific theoretical assessment of the buoyancy of gases in large fires overestimated the buoyancy effect.

During this period the design of the fire detection system for HGV shuttles was proceeding and this was to have an implication for fires on other rolling stock using the Tunnel, such as through freight trains. The approach to detecting a fire became a two-stage process. Initially, the idea of tunnel-mounted fire detection systems was proposed, and when it was found that it did not meet all the criteria for HGV shuttle fires a further onboard system was developed.

The in-tunnel fire detection system consists of multiple detection stations situated along the rail tunnels. Each station comprises three sets of detectors: ultraviolet and infrared flame detectors, ionisation and optical detectors, and carbon monoxide detectors. The logic controlling the detector and spacing of the stations had to be carefully considered and tunnel tests were undertaken to ensure that the system worked.

The onboard detection systems were mounted on the flatbed loading wagons and were based on an aspiration design which had to overcome the pressure fluctuations in the tunnel. Three companies were invited to develop this smoke detection system incorporating carbon monoxide sampling; all three chose an optical detection system. However, following the Hammerfest (Norway) and Buxton HGV tests, the carbon monoxide element was removed from the specification.

The overall design also had to take into account such matters as the fire resistance of the club car, the behaviour of the carrier wagons with a HGV fire, and the resistance of the catenary to fire.

The experimental programme, confirmation of the performance of the ventilation system and the ability of the detection system to identify a fire quickly, all of which protected the safety of passengers and any emergency services personnel in the tunnel system, persuaded the Safety Authority that the design and procedures were acceptable.

SERVICE TUNNEL TRANSPORT SYSTEM (STTS)

The service tunnel has four principal functions: to act as the main duct for ventilation; to carry cables, fire main and other equipment; to act as a safety refuge if passengers have to evacuate a stationary shuttle or train in the rail tunnels; and to provide a roadway for maintenance staff and emergency forces responding to a tunnel incident.

In respect of the last function, vehicles were designed on a basic chassis comprising a scrubbed diesel prime mover with solid tyres, a driving position at each end, and a guidance system using wires embedded in the roadway. While the chassis was common to maintenance vehicles and those used by fire, ambulance and police, there was an ability to mount a 'pod' which varied in design. (See Chapter 18 for further details.)

Two of the fire STTS vehicles, manned by firefighting personnel, permanently patrol the service tunnel, thereby ensuring a maximum 10 min delay in reaching any incident which may occur in the tunnel system. This ensures that a quick appraisal can be reported to the control centre and appropriate reinforcements sent on the other STTS vehicles dependent on the nature of the incident.

EQUIPMENT ROOMS

The tunnel system requires a number of areas in which electrical substations, communications equipment, control and computer equipment can be placed.

These areas are fully enclosed rooms with access only to the service tunnel. They are designed to contain any fire that may occur and are protected by detection and suppression systems, the latter using Halon 1301 to ensure a quick extinction of any fire which may start. The ventilation system for equipment rooms is interlinked to the detection and suppression equipment to ensure smoke does not penetrate the service tunnel.

While this chapter has described the long, complex process of designing the Channel Tunnel's innovative and comprehensive fire safety system, fire precautions are further detailed in the appropriate chapters, for example those on rolling stock (12–14), aerodynamics (15), service tunnel (16) and safety management (22).

17

Lighting

MICHAEL COWAN

When the design for the Channel Tunnel project started there was a general assumption that all three tunnels would be fully lit all the time; it was not long before people began to ask why. This led to initial studies of safety and ergonomic aspects, which indicated a strong operational preference, on safety grounds, for the rail tunnels to be virtually dark. Reliance on locomotive headlights avoids the issue of flicker from the tunnel lighting, and also helps drivers to recognise trackside signs. It should be remembered that railway signalling within the Eurotunnel operation is of the 'cab signal' type, where all advisory information to the driver is displayed on a video terminal in the cab.

The Concession calls for a full-length evacuation walkway along both rail tunnels, intersected by cross-passages every 375m, with fire-resisting doors. This feature facilitates the evacuation of passengers in an emergency situation in the event that the preferred method, of using the trains themselves, was not practicable. Consequently, development of the lighting criteria tended to centre on the lighting of this walkway, in consultation between Eurotunnel, TML's designers, the Safety Authority, the Maître d'Oeuvre and the various railway operators that would use the Tunnel. It was established that in normal operation the rail tunnels would be virtually dark, with marker lights at each cross-passage and small, unobtrusive guidance beacons using light-emitting diodes at 75m intervals in between combined with the push-button units that operate the main lighting. These give the drivers reassurance without being distracting. For the situation of an emergency calling for evacuation of passengers to the service tunnel, lighting on the walkways was specified at 20lux average (effective) with 0.5 uniformity, but with a power-failure mode of 10lux average (1lux minimum). This lighting is also available for safe access during maintenance 'possessions', supplemented by portable local lighting as required.

The service tunnel is more like a road tunnel, but the use of automatically guided transport vehicles makes it different from a normal road tunnel. The criterion of 20lux with good uniformity was also adopted here, but the operator has flexibility as to the extent of lighting use in normal conditions. One mode is for the guided vehicles (which have drivers) to drive on their

Figure 1. Rail tunnel luminaires mounted above walkway; piston relief duct on left. (Source: QA Photos)

headlights, with only those tunnel sections being illuminated where work is in progress as a safety measure. Again, transition lighting had to be considered at the portals, but not to the same extent as in a road tunnel, because there are air locks at each portal in which the vehicles have to stop briefly, since the service tunnel is also the main supply ventilation duct for the tunnels. The same philosophy of regular marker lights and a power-failure mode apply.

LUMINAIRES

The rail tunnels have over 13,000 luminaires, mounted in a single row above the walkway (Figure 1); they are designed for quick replacement of individual units, using a specially developed flexible plug-and-socket power system, to minimise maintenance time during possessions. The luminaires were developed specifically for the project, a process which included illumination tests in a 200m section of tunnel at the prototype stage, with a mockup walkway and full-scale mockups of the sides of the various types of train, to check their effect on illumination under evacuation conditions. All this was done in close collaboration with the main contractor TML, its designers EPDC and the

Balfour Beatty/Spie Batignolles joint venture. Great care was taken, by the French manufacturer Comatelec, with the optical design, which had already been tested in the Liège lighting laboratories of the manufacturer's Belgian associates. This enabled the required illumination and uniformity, both longitudinally and across the two-level walkway, to be achieved with 18W miniature fluorescent lamps, at spacings in the range 7.5–8m, with a mounting height of about 4m. The lamps have a correlated colour temperature of 3400K, and low-loss ballasts with electronic start are used. These are fed alternately from two different circuits, supplied from two different locations. Rail tunnel luminaire photometry is shown in Figure 2.

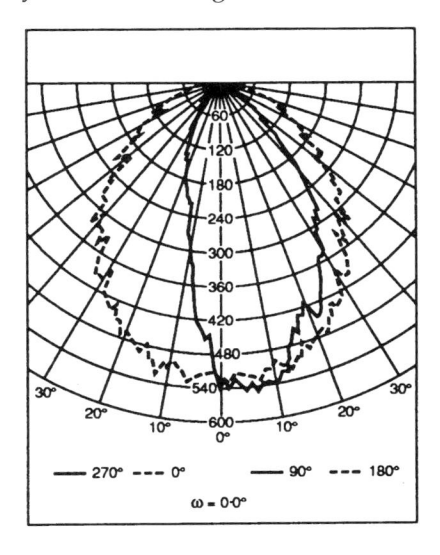

Figure 2. Photometry of rail tunnel luminaire. (Source: Comatelec SA)

SPECIAL APPLICATIONS

While the general standard of fire performance of all cables in the tunnel complex is lsf, tested to exacting underground railway standards, one of these two lighting circuits is in fire-resistant (micc) cable. Under fire conditions, service could be maintained with alternate luminaires, giving an absolute minimum of 1lux on the walkway. Owing to the uniqueness of the application, this figure was not derived directly from codes; however, careful consideration was given to British and French emergency lighting codes and international tunnel practice. The prime function of the installation is that of emergency lighting, in both the 1 and 20lux modes. Some early schemes were based on self-contained units, but it was decided that the intensity of railway operation called for minimisation of inspection and maintenance work within the tunnels. Use of self-contained units is limited to certain emergency markers only. In the service tunnel and adjacent spaces, a different design of luminaire is used, achieving similar illumination across the full roadway width from luminaires mounted at 2.5m, spaced 8.4–9m, and accounting for a further 7000 luminaires.

Another special application is the illumination of the enormous crossover caverns, as high as the nave of a cathedral, where the two rail tunnels meet each other so that the tracks can cross over, for single-line working and maintenance. In this case, the luminaires have 36W lamps and different characteristics; these other types of luminaire were all from the same supplier as for the rail tunnel.

The power supply to the tunnels is based on the principle of distributed redundancy. All safety-related systems can be fully supplied from either France or the UK via parallel feeders within all three tunnels. Notwithstanding the fact that these supplies are derived from 400kV grid substations on both sides, themselves interconnected by the cross-channel DC link, the highly safety-conscious approach used on the project resulted in diesel generator essential backup supply at each coast as well. The final circuit voltage adopted throughout the project is 400/230V, Eurotunnel thus leading the way in European voltage harmonisation.

Not only was the optical system carefully developed to suit the application; so was the cast aluminium enclosure with flat toughened glass front. The performance objective for the rail tunnels was specified as IP 65, but it was 'IP 65 with a difference', because a particular feature of the tunnel is the use of transverse piston relief ducts which interconnect the two railway tunnels. These act to reduce the air pressure buildup in front of the trains in a section 36km long between coastal shafts, and give rise to various patterns of air pressure fluctuation, so that luminaires (and other rail tunnel electrical equipment) had to be tested to achieve IP 65 in a simulated environment which represented extremes of pressure fluctuation associated only with abnormal conditions (± 30kPa). The luminaire design performed well in these tests; other special features included its aerodynamic form, tested at the von Karman laboratories in Brussels, since all rail tunnel installations had to meet prescribed aerodynamic criteria, allocated between them in the form of a drag budget. In addition, attention was given to the exterior finish (anodising and two-coat epoxy, with stainless steel fittings), to withstand the rail tunnel marine environment, with particular attention to internal support and fixing details to withstand vibrations.

There was an emphasis throughout on all safety-related equipment being specified and tested to stringent criteria to ensure functioning in the event of an emergency evacuation, stemming from the paramount importance given to fire and life safety throughout the project. Clearly, this has to extend to the control arrangements. While the basic evacuation principle is to get trains out of the tunnel without stopping, the lighting and its controls have to ensure operation in a situation where passengers have to be evacuated and transferred to the other rail tunnel, via the service tunnel safe haven. Local control is by illuminated push-button units at approximately 75m intervals, any one of which will activate three 750m long sections of lighting in both the affected tunnel and the adjacent service tunnel, thus ensuring that the full length of walkway alongside the train is illuminated, no matter where it stops. However, the normal mode of operation is for these sections to be similarly energised by remote control from the main control centre in the Folkestone Terminal or from the control centre in the Calais Terminal. This is managed by the Engineering Management System, which communicates via the fibre-optic data transmission system.

For details of exterior lighting, see Chapter 9.

18

Service tunnel and vehicles

PETER SEMMENS AND YVES MACHEFERT-TASSIN

The service tunnel has a number of functions. It is used for access to the technical equipment located in the cross-passages and in the equipment rooms along it. It also provides fresh-air ventilation for the rail tunnels, and acts as a 50km-long safe haven if a shuttle or train has to be evacuated. The service tunnel transport system, or STTS, enables staff to reach every point in the Tunnel, if necessary at short notice. When the Tunnel was being designed, it was initially intended to use a narrow-gauge railway system for access. However, although this would have provided guidance as well as speed, the system lacked the flexibility of a road accommodating rubber-tyred vehicles. Accuracy of lateral positioning in a small tunnel is essential and, after extensive testing, it was decided to use specialised rubber-tyred vehicles whose steering, and therefore their position in the Tunnel, can be controlled

Figure 1. LADOG (left) and STTS vehicle (right) in the service tunnel. (Source: QA Photos)

by a buried-wire guidance system. The initial fleet of purpose-built, double-ended STTS vehicles was subsequently supplemented with lighter vehicles, known as LADOGs, derived from commercial runabouts used extensively throughout continental Europe. Both types of vehicle are shown in Figure 1.

STTS VEHICLES

Twenty-four STTS vehicles have been constructed. Each consists of a framework that connects a cab and power unit at each end, enabling it to carry a variety of different 'pods'. Their primary role is maintenance, but they are also used for firefighting and other emergency purposes; the pods are colour-coded to indicate their function (yellow, maintenance; red, firefighting; white with 'red cross', ambulance). Special pods for each different type of service are inserted from the side, after blocking up the vehicle to prevent its suspension sinking as the weight is transferred (Figure 2). The payload is 2.5–3 tonnes.

Figure 2. Inserting a pod on an STTS vehicle. (Source: QA Photos)

The STTS vehicles are driven from the leading end. They are too long to be turned in the Tunnel, so when they need to reverse the driver changes ends after centralising and locking the steering and shutting the motor down. The power unit at the new leading end is activated, and the vehicle can then be driven back to its starting point.

Two pairs of continuous wires run under the floor of the service tunnel. Sensors on the vehicles enables them to lock on to the left-hand wire of the pair so that the driver does not have to steer. In this mode they can travel up to 80km/h. If they are being steered manually, speed must be kept below 50km/h.

Each vehicle has two Daimler-Benz OM 602 diesel motors (as used on the Mercedes 190), individually rated at 62kW (83hp) at 3400rev/min. Each motor drives the nearer axle through a Daimler-Benz W4A 028 automatic transmission; only one is use at any time. The original intention was to provide an electric drive as well, but the ventilation in the service tunnel has proved to be so good that there is no need for this added complication.

Each rubber-tyred wheel is fitted with an internal 'run-flat' disc, enabling it to continue if it suffers a puncture. The design is derived from the pneumatic-tyred railcars and coaches that operated extensively between the two world wars on the French railways. (An example is on display in the railway museum at Mulhouse in eastern France.)

Two STTS vehicles are on duty in the Tunnel for firefighting purposes at all times, staff also dealing with routine duties during their shifts. A backup vehicle is constantly available.

LIGHT SERVICE TUNNEL VEHICLES

The need for these vehicles became apparent while the Tunnel was being built. A larger fleet of vehicles enables more efficient use to be made of the staff who maintain the interior of the Tunnel and the wide variety of equipment it contains. Without the light vehicles, the STTS vehicles would have to operate a 'bus service' at set intervals, delivering a succession of people and equipment to different points in the Tunnel. Having completed a particular job, the person concerned would have to wait for the next return vehicle.

A fleet of 15 LADOGS was purchased to circumvent this restraint and to assist maintenance and other work in the terminal areas. Built by Nordrach at its factory in the Black Forest, they are used extensively throughout Europe. The LADOGs have a short wheelbase, with hydrostatic drives to each wheel; all four wheels can be steered, the minimum turning circle being only 3.4m. This enables them to make a two-point turn in the width of the service tunnel, the first move being in reverse. However, most drivers generally carry out three-point turns, as they do in the open. For normal driving, the rear-wheel steering is locked; power is split between the four wheels to avoid one end getting bogged down on soft ground in the open. Experience has shown that a differential lock is not necessary, even when the vehicle is operating in deep snow.

Each LADOG is powered by a 40kW (54hp) Volkswagen diesel motor, with separate injection chambers for each cylinder. This gives a cleaner fuel burn at a lower temperature, helped by the reduced compression ratio. The power output for the given swept volume is also kept low, and the motor works at relatively modest rotational speeds. Maximum speed in the Tunnel is 50km/h. Unlike the STTS vehicles, the steering of the LADOGs cannot be locked to the buried guide wire.

Their hydrostatic power arrangement enables LADOGs to be fitted with auxiliary equipment powered by the same means. Some are provided with rotating sweeping brushes, washers or vacuum drain cleaners, and all have attachments for snowplough. While the shuttles themselves are not affected by severe weather, it is necessary to keep the roads and platforms in the terminal areas clear of snow. A light fork-lift attachment is also available at the front of the vehicle.

LADOGs are not provided with run-flat wheels, unlike the large STTS vehicles. However, each carries a foam-injection cylinder to reinflate a punctured tyre, enabling them to leave the Tunnel. Spare tyres are available, for them and for the large STTS vehicles, at the base just outside each portal.

The LADOG fleet consists of two different types of vehicle. The 18 intended primarily for use inside the service tunnel are slightly shorter: model number AS129, their lifting cabs for access to the motors can be raised inside the Tunnel. A variety of pods of up to 1 tonne in weight can be accommodated on the rear of the vehicles, including a personnel carrier. These are jacked off (Figure 3) and supported on legs when not in use. The weight of an empty

Figure 3. Offloading a pod from a LADOG. (Source: QA Photos)

LADOG is 2.5 tonnes, half the weight of a loaded STTS vehicle. The LADOGs used in the Tunnel are 3.3m long, 1.25m wide and 2.05m high.

Since road vehicles are driven on opposite sides of the road in France and the UK, a standard rule of the road had to be adopted for the road vehicles in the service tunnel. It was agreed that they should drive on the left, and the drivers therefore sit on the right. Because of the slight danger that a continental driver might subconsciously steer to the right when another LADOG was approaching, the Safety Authority required warning sensors to be fitted to each vehicle. These alert the driver if a vehicle strays significantly across the Tunnel to the right.

IV

TERMINALS

19

Terminal design and construction

MARTIN STEARMAN

The design and construction of the terminal sites at Cheriton (Folkestone, Figure 1) and Coquelles (Calais, Figure 2 overleaf) included certain straightforward tasks but also others where a new approach or design solution was required. This chapter identifies only some of the more interesting features of the design, due either to the uniqueness of the solution or to the scale of the task, and describes the way in which decisions in one area inter-relate with those in another. The view presented is not that of the designer or contractor, but of the client's representative, acting on behalf of the recipient of a complex design and building contract.

TERMINAL DESIGN

Function and corporate identity

Eurotunnel's brief for the design of the project required the contractor and the design subconsultants to undertake a functional development of the original

Figure 1. Folkestone Terminal: aerial view, 1994. (Source: QA Photos)

Figure 2. Calais Terminal: aerial view, 1994. (Source: QA Photos)

1970s' schemes, but in such a way as to make the operation of the Channel Tunnel system and the users' experience consistent with the required corporate image of Eurotunnel. The commercial requirement was for the terminals' design-level operational throughput, in tourist and freight traffic, to be achieved.

Terminal layout

The functional design requirements of the terminals were concise: to take vehicles from the motorway system and load them onto shuttle trains. The scale of this operation (700 cars and 113 heavy goods vehicles per hour), the innovative nature of the operation and the time limits imposed were more onerous. Taking vehicles directly from a motorway with the minimum of interruption directly onto an enclosed railway wagon in less than 15 min was significantly more demanding than any such operation elsewhere in the world.

The absence of similar facilities elsewhere was a particular problem given the need to establish the design quality of the terminals given the 'design and build' nature of the contract between Eurotunnel and the contractor. As a consequence, the concept of 'reference buildings' was proposed by advisers to Eurotunnel at an early date, identifying Terminal 4 at Heathrow and Charles de Gaulle Airport at Roissy near Paris as the comparative buildings. However, neither of these 'reference buildings' properly reflected the unique function of the terminals.

On the UK site the first major issue was the form of the connection with the M20 motorway. As a result of representations to the Select Committee of the House of Commons, a change was made from the route of approach between the two villages of Newington and Peene to one on the southern alignment of the Continental main line. This required a major design development exercise of the frontier controls area of the terminal.

At an early point it was realised that the available area within the terminal site, constrained as it was by the motorway and escarp scope on the North Downs, was extremely limited, both in length and breadth. This was to create significant technical difficulties later and prolong the design period because a change in one area would impact on an adjacent one. The constraints in space resulted in a very intricate layout of road and facilities, such that, when it was found necessary to modify the radius of a road kerb, a knock-on effect was inevitably caused in some adjacent area of the layout. In these circumstances the final definition of the layout required exhaustive reworkings of the design process.

The basic concept of terminal organisation was that within each country the UK and French frontier controls would be juxtaposed on entry to the system, thus providing a free exit, i.e. having left the shuttle on arrival in the destination country, the traveller would be free to drive directly onto the national motorway system without further checks or delays. It was originally assumed that duty free facilities would be between the UK and French controls or vice versa. However, consideration of the terminal design in process terms showed that visiting the amenity buildings should really be regarded as an 'off-line' activity.

During sensitive negotiations with the relevant authorities it was agreed that access to the amenity facilities could take place after the tolls but prior to the frontier controls system being entered. This arrangement could only be achieved by an undertaking being given that once people were in the system they would not be free to leave unless explicitly allowed to do so, other than through the rail system or under emergency conditions. Complicated discussions as to what was 'landside' and resulted. An effect on the terminal operation was that security could be maintained more easily, but that the terminals could not be opened to visitors.

In France, achieving the suitable layout for the Terminal was somewhat easier, given that sufficient site areas were available for the tolls, amenities, security controls, allocation and loading 'process' to be taken in relatively straight lines.

In the UK it was necessary to organise road layout and routeing in a slightly more convoluted fashion with routes folding back on each other to achieve the required distance so that processing could take place. However, it was imperative that the customers using the system were not aware that they were doubling back, and the time taken to actually pass through the system was minimised.

Traffic management – organisation and security

At the same time as terminal layout development work was going on, the traffic management system and the vehicle allocation system were being designed to ensure the right vehicle ended up on the correct shuttle. To achieve this required that each type of vehicle follow a predetermined route through the allocation area and into the correct lane in the holding area where vehicles are gathered in line before proceeding in 'crocodile' fashion across the overbridges onto the correct shuttle. The nature of the rolling stock with three forms of shuttles, i.e. double-deck tourist, single-deck tourist and freight, required that each type of vehicle be identified and allocated to the appropriate location for efficient loading, security and safety. Terminal traffic management is explained elsewhere (Chapter 20) but the organisation of the road system had to facilitate the 'processing' of the vehicles (Figures 3–5).

After the final road layout had been determined, it was necessary to introduce a major security inspection facility on the freight route as part of the processing of freight vehicles. The solution found underlines the limited nature of the UK site. This facility could only be introduced by creating a complete circular loop out of the normal route, with the vehicle taken out and reintroduced in the same position, as there was insufficient length of road available between the tolls and the loading platforms simply to add on a new sequence in the processing of the vehicles.

Figure 3. Cars at toll plaza, Folkestone Terminal. (Source: QA Photos)

TERMINAL DESIGN AND CONSTRUCTION 259

Figure 4. Overview of traffic flow: allocation areas to platforms, Folkestone Terminal. (Source: QA Photos)

Figure 5. Tourist vehicle toll and control system, Calais Terminal. (Source: QA Photos)

Figure 6. Platforms and overbridges, Folkestone Terminal. (Source: QA Photos)

Platform areas and overbridges

An area where the constraints on the UK site are clearly indicated is in the platforms and overbridges area (Figure 6): different forms of structures were adopted in the two terminals. Early in the design process various different platform arrangements were investigated.

To achieve the design throughput of cars (90 per shuttle), the shuttles were designed to carry cars in a double-deck arrangement.

Some early design studies for the platform area included both low-level and high-level platforms. Such a configuration failed to provide a sufficient level of flexibility in terms of which shuttles could use which platforms. The problem was solved by having ramps inside the shuttles to provide access to the top decks, so the platforms could all be at the same level. The final flexible and simple platform arrangement was to have single tracks separated by double-sided platforms. Achieving this apparently straightforward arrangement took a considerable amount of design effort on behalf of both the contractor and the client, impacting directly upon the design of the shuttles themselves, which were in turn the design responsibility of a separate subcontractor.

The interactive and iterative nature of the design process involved in the platform area clearly indicates the multidisciplinary nature of that design process, involving rail, road, civil, structural systems, operational engineering and even architectural and aesthetic considerations.

In France the traffic access to the platforms takes place by means of four bridges at right angles to the railway tracks. In the UK a parallelogram format had to be devised between the entry fan of rail lines leading into the platform block and the corresponding exit fan, where the trains leave the platforms before entering the tunnel portal.

Whilst the complex longitudinal geometry was being established, with the platform and the overbridges being forced into the available spaces between the railway fans at each end, a detailed examination of the physical constraints across the UK terminal site in turn affected the design of structures on the southern perimeter and established the need for additional land to the north of the site.

The relative levels above datum of both the site itself and the connections into the adjacent motorway, railway systems, and the vertical and horizontal alignment of the Channel Tunnel entrance were all critical considerations. The vertical alignment of the rail layout of the terminal was influenced by critical considerations at the portal, the M20 crossing and the limiting level of the rail track on the northernmost part of the terminal at the foot of the scarp slope where there was a need to limit the extent of excavation.

At the western end of the site a series of design and engineering studies were necessary to fix the finished construction levels and intersection of new connection from the British Rail line into the terminal, leading ultimately to the Tunnel. This required the realignment of the existing A20 road in a cutting and the construction of approach and exit roads from the terminal. These road decisions were affected by the need within the terminal to find the correct rail alignment and geometry for the loop taken by the shuttle trains as they passed beneath the main line and returned, facing the way they had arrived, into the platform area.

At the eastern end of the site the precise location of the entrance portal of the Tunnel was constrained both horizontally and vertically in geotechnical terms, because of the need to construct the Tunnel through the base of Castle Hill, utilizing the layer of gault clay before moving into the layer of unweathered chalk marl (see Chapter 2).

The vertical alignment of the rail line was also critical because of the need to minimise the steepness of the gradient of the exit from the cross-channel section of the Tunnel. As the degree of slope increases so does the power requirement for the locomotives or the speed drops, thus preventing the through trains of BR/SNCF from maintaining the time schedules necessary to achieve the desired journey times in competition with the airlines.

After all these complex considerations of geometry and site levels, the organisation of traffic routeing and traffic management, the resulting terminal layouts have succeeded in satisfying the original Eurotunnel criteria, not only in functional but also in design terms.

CORPORATE IDENTITY

Corporate identity design considerations were discussed at an early stage and were formative in the design development of the project.

Early work on the public response to, and potential use of, the Tunnel system indicated that a proportion of potential passengers might find the idea of travelling in the Tunnel daunting. In order to help minimise this effect and therefore maximise potential usage, a corporate design strategy was developed governing the design of all facilities, buildings and layout with the intention of enhancing confidence in the system.

'Key word descriptors' were identified which could be applied to the design of all elements of the project. The key words included definitions of the aspirations of the system, i.e. that it would be *rapid, modern, stimulating,* but equally that it must also be felt to be *sympathetic, inviting,* and, above all, that the organisation should be *understandable* and the Channel Tunnel as a whole seen to be *safe*.

After this early work by BDP and ADP as the design subconsultants to TML, Wolff Olins and ADSA were brought in as strategic identity advisors to Eurotunnel. A corporate identity working party was established to examine all design proposals and to provide guidance to specialist designers. This working party remained in place in various forms throughout the design and implementation of the project.

The impact of these design considerations can be seen not only from the organisational layout of the terminals, as described above, but also in the approach to the design of all the sign systems, landscaping, buildings, access to the terminals and in the rolling stock and operational vehicles. It was recognised at an early date that the potential for the terminals to become complex, busy and visually disruptive had to be reduced. Consequently, a design strategy which was developed required that all buildings, whilst being functional and efficient, should be recessive and blend into the background, unless they have a specific need to be dominant, such as the control centre and the passenger terminal (amenity) building on the UK side. An early internal Eurotunnel paper encapsulated the aesthetic requirement for the Folkestone

Figure 7. Passenger terminal building, Folkestone Terminal. (Source: QA Photos)

Terminal buildings as being 'white pavilions set in the green landscape' (Figure 7).

Taking a lead from Disney, a concept was developed of a 'Eurotunnelworld', clearly defined as one entered the system from the motorway either in the UK or France. This concept should be reinforced and maintained throughout the system. A key element in achieving this objective would be the signing and information systems.

Also important in ensuring a consistent corporate identity between the terminals was that there should be a limited range of building materials and a uniform colour palette. The two contractors, each undertaking one terminal, were required to produce design strategies for the terminals based upon the key word descriptors.

Efforts to secure joint working between the UK architects (BDP) and the French (ADP) achieved limited success apart from the early work, due partly to pressures from their respective contractor employers. Consequently, the overall Design Strategy Concept was devised to allow the architects to develop ideas in their own fashion but within an agreed basic framework. The design freedom that this created, operating within a framework, is illustrated by the form proposed for the canopies at the two terminals (Figures 8 and 9). The flat landscape of the Calais Terminal demands vertical emphasis. ADP developed a reversed square umbrella strongly mounted on single columns. In the UK, BDP developed a lighter, more delicate space frame format reflecting the need to express a horizontal emphasis in contrast to the dominance of the vertical scarp slope which forms a background to the site. Whilst these two approaches seem markedly different, it was intended that

Figure 8. Canopies over frontier control, Calais Terminal. (Source: QA Photos)

Figure 9. Canopies over frontier control, Folkestone Terminal. (Source: QA Photos)

the user passing through the system would be reassured by the continuity provided by the use of simple forms and similar white, grey and silver materials.

The design of the civil engineering structures at the Folkestone Terminal also reflects the psychological requirements set out in the Design Strategy. Here the key words *safety, security* are paramount. The structures are seen to be firmly anchored into the ground and the areas of exposed concrete structure are modified to relate to the human scale by means of a horizontal banding detail.

A fundamental element in the visual continuity between the UK and France was the achievement of a Eurotunnel signing system. To achieve this required considerable effort and ingenuity by many people working towards a common objective. Initially the respective official sign systems of the UK and France

Figure 10. Traffic signs at Folkestone Terminal. (Source: QA Photos)

were the only sign format that the Intergovernmental Commision (IGC) would allow. A Eurotunnel-specific sign system was proposed which, whilst using the mandatory official road signs of the UK and France in the respective terminals, was developed via the informative signs to effect a transition between the two countries. The design consultants ADSA were particularly instrumental in this process, developing a rigorous if, in the event, slightly subjective analytical system using a dominant and a subordinate language, i.e. French/English or English/French, depending on the country of application. The net effect is a system unique to Eurotunnel which fits comfortably with the developing aesthetic for European sign systems (Figure 10).

ENGINEERING CHALLENGES

Site fill

As a result of the variations in the existing ground levels, the finished ground level of the Folkestone Terminal site had to be raised by 12m in some places. Originally the material to be used for site fill was to be minestone from the Kent collieries and chalk spoil which was due to be excavated from both the land and the marine tunnels. Import of the minestone carried with it an inevitable environmental impact resulting from the number of vehicles that would be required to move such vast volumes of materials. An alternative was developed by Westminster Dredging to provide sand collected from the Goodwin Sands off Deal and delivered by an overland pipe from the coast at Hythe, saving half a million lorry movements. Subsequent to this decision to use marine-dredged sand was the final acceptance of the proposal to contain all the UK excavated tunnel spoil at Shakespeare Cliff in a series of lagoons behind a sea wall, thus creating a new 'platform' at the foot of the cliffs (see Chapter 5).

Loop tunnel

Within the terminals there are a number of structures which, by virtue of their size, would be regarded as significant engineering projects in their own right had they been built anywhere else. The most interesting of these at the Folkestone Terminal is the loop tunnel. Built as a concrete box on a 'cut-and-cover' basis, it is 1100m long by 20m wide and 8m high at a tight radius. At the Calais Terminal the rail loop to turn the shuttles around for the return journey is at ground level. However, in the UK the physical constraints of the site, together with the need to safeguard the immediate environment of people living in the villages of Newington and Peene, required that the 'loop' should be below ground as far as possible. It was, in fact, constructed only partly underground but with a heavily planted earth mound placed over the top, thus providing a visual and acoustic screen between the area of the terminal in the middle of the loop and the residential properties immediately outside the site perimeter.

A number of different structural solutions were investigated because not only did the ground conditions vary over the length of the loop tunnel, but the southern end construction, requiring a 9m deep cutting, was immediately next to a listed building that had to be retained. At the northern end, construction passed just south of the toe of a potential slip zone to an ancient landslip at Danton Pinch.

As a result of the extensive safety considerations applied to the whole of the Channel Tunnel project, the loop tunnel had to be designed to enable people to escape from the shuttles should there be an emergency. Consequently, the loop tunnel was constructed wider than required for the shuttles, to allow for fire exit routes between tracks, and escape stairs are provided at the midpoint of the 1100m tunnel with smoke vents located at the intermediate positions.

As this example shows, there are a number of areas on this project which combined the characteristics of rail, road and pedestrian installations. Appropriate design criteria had to be established and agreed with the IGC before the project could be completed.

Brickpit structure

Between the M20 and the new Continental main line lay the main Central Electricity Generating Board cross-channel cable. Alternative methods were investigated of providing structures over or beside the cable, including construction of a landscaped slope with a tunnel inserted over the line of the cables, in order to avoid risk of disruption of the power supply.

However, a reinforced earth retaining structure, one of three at the Folkestone Terminal and the biggest of its kind in the UK (14m high) was the selected option to carry the main line and the shuttle arrival and departure lines. It is the first reinforced earth retaining structure to carry a main-line railway in the UK. Reinforced earth is a method of constructing embankments which uses the weight of the soil, acting on horizontal metal strips attached to the embankment wall slabs, to maintain stability.

Whilst a proposal for a landscaped earth slope with the cross-channel cable enclosed in a tunnel was clearly preferable in environmental terms, there were still concerns over whether the construction work could be carried out without disturbing the cable, which would have serious financial repercussions on the contractors. Consequently, the reinforced earth option selected as the best technical and, in fact, the cheapest solution but because of the difficulties in ameliorating the visual impact of the structure it was most problematic proposal on which to secure planning approval from the local planning authority.

Extremely limited space between the M20 and southernmost rail line emphasised the criticality of the limited overall width of the site, an old brickpit. Additional land was secured north of the terminal in order to accommodate landscaping. This enabled the civil engineering to be extended as far north as possible, whilst providing the tree planting promised earlier, but outside the original site boundary.

Bigginswood

Part of the Folkestone Terminal site was occupied by Bigginswood, a remnant of seminatural ancient woodland. Whilst it was not officially designated as a Site of Special Scientific Interest, it aroused some public attention.

The high profile of the Channel Tunnel project resulted in long discussions and negotiations as to the future of Bigginswood. TML devised a rescue strategy, whereby a hectare of the land surface occupied by the woodland would be removed and relocated to an area of the terminal rendered similar to that of the original site. Samples of vegetation were collected, maintained offsite and later reintroduced into the new location as recommended in the 1985 Environmental Impact Assessment (see below). The successful outcome of the experiment owed much to the combination of research, adherence to carefully established preparation, and working methods using standard site machinery.

UK PLANNING PROCESS, CONSULTATION AND ENVIRONMENT

The means of securing formal approval for the project did not follow the usual public inquiry route. The Channel Tunnel Act gave outline planning permission for the construction of the UK Channel Tunnel facilities in terms of location and scope but in reality imposed greater demands on the developer

(Eurotunnel) than under established planning law. The local planning authority had additional powers regarding position of facilities in the terminal, and as to the method of construction and volume of materials that were to be employed.

A Joint Consents Team was set up by TML and Eurotunnel, staffed by experienced town planners, who were given a remit to undertake liaison discussions with the relevant local planning authorities on behalf of the project to facilitate approvals. An agreement was reached between the local planning authorities and the promoter Eurotunnel, which resulted in the addition to the Channel Tunnel Bill of a clause providing for special town and country planning arrangements; the signing of a joint Memorandum governing the operation of this clause; and a series of government conditions, assurances and undertakings relating to this clause. The whole process was monitored by the Joint Consultative Agreement, which was chaired by a government minister and included representatives of central and local government and of the concessionaires. These arrangements also covered areas beyond the terminal, such as Shakespeare Cliff, Dollands Moor and Ashford. Any political decisions required would be taken at the Joint Consultative Committee. The Royal Fine Arts Commission was also consulted about the design of the terminal.

In anticipation of future EC requirements an Environmental Impact Assessment (EIA) was prepared in 1985. The nature and extent of the studies involved were contributory to the securing of the Concession to build the Channel Tunnel. The EIA included recommendations on minimising the physical impact of the development as well as identifying design parameters to mitigate the effect of the project on the local population. It was recognised by both TML and Eurotunnel that, by respecting the design parameters identified, potential criticism of the project would be minimised.

A pro-active approach towards public consultation, environmental and design issues was developed early in the project, focused on the Eurotunnel Exhibition Centres in Folkestone and Calais. An extensive range of societies and organisations were identified as interested parties and involved in a consultation programme. In addition to the normal functioning of the Exhibition Centre, special events were organised, such as 'open days' for the local residents and technical briefings given within the framework of an Annual Environmental Forum to members of interested groups. The Annual Environmental Forum encapsulated the open, informative and interactive nature of the public consultation philosophy and programme adopted by Eurotunnel. Arrangements included formal annual visits to the construction site for consultee groups.

Eurotunnel, through the Exhibition Centres in both France and the UK, made information, technical papers and documents available to both general and specialist enquiries. This open approach and availability of information was, it is believed, significant in raising the level of debate concerning the project and was a major contribution to the relatively low level of objections made about the project during the implementation work.

At the time of the Select Committee process, taking in both the House of Commons and the Lords over 6000 individual objections to the project were lodged. The effectiveness of the pro-active approach to public consultation is demonstrated by the fact that there were no residual objections to the project on environmental grounds: they were addressed in the course of the consultation period.

Listed buildings: removal and relocation

The construction of the Folkestone Terminal had a direct impact on three listed buildings. The road access from the M20, following the realignment to the southern route mentioned above, caused the removal of Mill House and Stone Farm. These buildings were the subject of survey and archaeological study, and were finally dismantled and placed in store for subsequent rebuilding.

Longport House, adjacent to the service entrance to the terminal off the A20, had originally been intended to be retained and construction work for the loop tunnel was specifically modified to protect the ancient building. Ultimately, however, detailed consideration of the most appropriate method to refurbish the building to bring it to an acceptable environmental standard and to find an alternative site for a new police building, permission was given to remove it.

Following an even more extensive archaeological study than that carried out on and Mill House and Stone Farm, arrangements were made with the Canterbury Archaeological Trust and the Weald and Downland Museum, as a result of which Longport House was rebuilt at the Museum's site in Singleton, Sussex, to become the main entrance and visitor facility. Unique information has been discovered on the original construction and changes that had occurred to the building over time.

20

Terminal traffic management

JOHN DAVIES

The management of road traffic through the railway shuttle terminal on the UK side of the Tunnel is similar to that in France and is conveniently divided into three components:

- **Tolls**: the need not only to collect money from customers who simply turn up at tolls, but also to check and process a wide range of 'authority to travel' documents such as prepaid tickets, special concession tickets for shareholders, cards and tickets for freight companies and others who have an account with Eurotunnel.
- **Vehicle management**: the organisation of a continuous stream of traffic arriving at the terminal into groups or packets of vehicles 'tailored' to fit into the discontinuous system of shuttles departing at varying intervals from the platform zone.
- **Signs**: a fundamental part of terminal traffic management (TTM) was the need to produce road signs which would meet the triple requirements of informing and warning customers, harmonisation with a system-wide signing and design policy, and compatibility with the national traffic sign system of the two separate countries.

The contract for provision of TTM was awarded by TML to the Unitraffic consortium comprising among others: SEMA (overall contract management and toll system); Logica (management software); Siemens Plessey (equipment and installation); and SES (signs).

This chapter is divided into three sections, one for each of the above components. Each section deals with the historical development of the system through the contract and implementation phase with TML, and outlines the key characteristics of the systems as installed at the start of operation.

See chapter 19 for the design and construction of the terminals themselves.

TOLL SYSTEM

Development

In the original Channel Tunnel contract, the general description of the tolling arrangements was included as part of the civil engineering contract under the terminal and fixed equipment heading. Although in the description of the UK terminal only a general mention was made of a toll system, in the section on the French terminal a system was described in great detail (number of different tariff categories, etc.).

A major source of difficulty was that the toll system described was evidently based upon the sort of systems that are installed throughout the French autoroute network. It visualised similar segregation of vehicles into tariff categories (even including the measurement of the height of the vehicle bonnet at the front axle!) and similar payment philosophies (stored value tickets, use of credit cards, etc.). This in turn reflected a fundamental difference in the view of the entire Channel Tunnel system between the French, who see it as an extension of their motorway network, similar to the Mont Blanc or Fréjus Tunnels, and the British, for whom the system is a new player in the fiercely contested cross-channel travel market and, as such, would have to be competitive with the ferries.

The trend over recent years in the travel market has, in fact, led to a compromise somewhere between these two extreme views. The ferries are moving away from reservations and rigid compartmentalisation of vehicles by length towards a single price per vehicle and turn-up-and-go. Similarly, a motorway toll system appropriate for large volumes of regular traffic and relatively small tariffs (£5–10) cannot be directly translated to a transport system which the majority of the customers will use only once a year and where the price of a return ticket can be more than £200.

As the toll system and TTM as a whole moved through the various stages of outline and detailed design, the original system outlined in the construction contract was amended in discussions between TML and Eurotunnel. In particular, flexibility was sought with a widening of the possible combination of vehicle categories for tolling (even if only a very limited number would be used at any time) and the expansion of tariff bands.

However, this slight movement away from the construction contract concept was unable to keep pace with the development of Eurotunnel's commercial policies. TML, with Eurotunnel's agreement, started the tendering process for the 'base case' toll system in May 1990. Subsequently an understanding was reached whereby, once the contract had been awarded, Eurotunnel was at liberty to seek a separate contract for modifications and enhancements to this base case system. Eurotunnel's problem was to ensure that the toll system, in addition to handling pay-at-tolls customers, could cope with regular customers travelling on a credit account basis, processing travel privileges to which more than half a million shareholders were entitled, and prepurchased tickets issued through travel agents or directly from Eurotunnel.

Thus, the TML base case system was developed and installed to provide an integrated equipment system meeting contractual performance and reliability standards. In parallel, but separately, Eurotunnel awarded a contract to SEMA for the enhanced system which customers will experience when they use the Tunnel. Currently an interim system is coping with freight and initial levels of tourist (i.e. passenger-vehicle) traffic.

Operational system

The toll systems are installed in three separate toll plazas:

- French tourist: 12 lanes – tourist vehicles only
- French freight: 3 lanes – freight vehicles only
- UK: 14 lanes – taking both types of traffic but with freight grouped on one side (Figure 1).

Control is achieved through a hierarchy of computers:

- toll lane computer or toll collector terminal (TCT) which controls the selling of tickets to turn-up-and-go customers and checks the validity of prepaid tickets or special cards/tickets for Eurotunnel account holders
- toll station computer (TSC) which controls and supervises the traffic in the lanes, under the management of the toll supervisor

Figure 1. Approach to toll plaza, Folkestone Terminal. (Source: QA Photos)

Figure 2. Toll lanes with variable-message signs, Calais Terminal. (Source: QA Photos)

- main toll computer (MTC) which processes toll and traffic data and exchanges this with the overall Eurotunnel management information system (MIS).

TOLL COLLECTOR TERMINAL

All the toll lanes are designed to be operated in manual mode, i.e. with a toll operator in the toll booth adjacent to the toll lane (Figure 2). Vehicles approaching the toll plaza are guided to those lanes that are open by the normal green arrow (open), red cross (closed) signs above each lane. In addition, variable-message rotating prism signs are installed above each lane (in a group of four) which can indicate further levels of segregation according to vehicle type (lorry, coach, car, car and tow etc.), manned or automatic lane, and for customers with prepaid tickets. Toll lanes are arranged either side of a double side-by-side toll booth so that right-hand drive vehicles will pass to the left and left-hand drive vehicles to the right, enabling both British and Continental drivers to carry out the transaction directly with the toll operator.

Between the entry to the toll lane and the toll booth, vehicles pass two height limit detectors:

- 4.20m: vehicles over this height must be rejected at the entrance before they have undertaken the toll transaction since they cannot be carried in the shuttle wagon
- 1.85m: vehicles over this height travelling on tourist shuttles must travel in the single-deck wagons.

At the toll booth the operator will input the characteristics of the vehicle into the terminal according to the Eurotunnel tariff categorisation. If the driver has a prepaid ticket or similar authority to travel the ticket with a magnetic strip is read by a ticket decoder which compares the vehicles and journey details encoded on the ticket with those entered by the toll operator. If there is agreement the customer is accepted; if not the customer is asked to pay more or is given a refund depending on the circumstances. For customers not already in possession of a ticket, the total fare they have to pay for their journey is calculated and displayed to the customer via an external dot matrix display. The customer may pay in either British or French currency and by a variety of methods (cash, credit card, cheque or traveller's cheque).

TOLL STATION COMPUTER

The toll supervisors are aided in their control of the activity on the toll plaza by the TSC. In particular, the TSC oversees the traffic (including rejected vehicles) and the payments in each toll lane during a toll operator's shift. At the end of the shift, the operators must reconcile their takings (including any float they may have received at the start of the shift) with the results calculated by the TSC. This is done under the control of the toll supervisor. Once the toll operator's takings are agreed, the system passes the information on to the MTC and the takings are despatched securely to the bank.

To safeguard the flow of cash and other items of value, a pneumatic fund transfer system is fitted between the toll control building and each booth. The initial and intermediate cash floats are managed by the TSC. The operator's takings are despatched to a separate, safe storage location prior to final reconciliation.

The TSC supervises the equipment in each toll booth and the flow of traffic in real time. The system also includes some lanes with automatic ticket readers and lanes able to accept vehicles with special 'dynamic' encoded tags. When these are used, they are controlled directly by the TSC from the remote location of the toll control building.

MAIN TOLL COMPUTER

The main function of the MTC is to manage the transfer of data between the toll system and the management system computers (especially MIS). Its tasks include:

- downloading of credit card blacklists from the banks/card merchant acquirers to the individual toll booths (TCTs)
- management of tariff tables and changes in tariff bands
- transmission of summary traffic and payment data to MIS for use in high-level management summaries
- linking to the traffic control and supervision computer (see below) to exchange traffic data and to react to any emergency lane closure or opening instructions coming from the traffic supervisor.

VEHICLE MANAGEMENT

Development

The second major component of the TTM is the vehicle management system itself which must take a free-flowing stream of traffic from the motorway and segregate it progressively until the correct number of the correct type of vehicle is available for loading into the special railway wagons that make up the shuttle train. Between the tolls at the entrance to the terminal and the shuttles in the platform zone it is necessary to guide and control the vehicles through a terminal comprising:

- a passenger terminal building containing services such as toilets, restaurants, shops and duty free sales
- security and frontier control checks for both countries (including the possibility of extended checks)
- a marshalling and allocation area in which vehicles are arranged into groups that match the composition of the shuttles
- an arrangement of bridges and ramps taking vehicles from the road system on the terminal down to the railway loading platforms.

In addition, it is of course necessary to direct customers arriving by shuttle away from the platforms and out of the terminals with the minimum risk of confusion. The fact that all frontier controls had already been completed at the departure terminal greatly assisted the design and operation of these arrivals.

Two factors played a major part in the organisation of the terminals and the traffic management between the original concept and the final realisation: the arrangement of the passenger terminal building and frontier controls, and the design of the shuttles and the regulations that arose from loading vehicles into these shuttle wagons.

The first factor was resolved in agreement with the two governments and the Intergovernmental Commission (IGC) relatively quickly. The original design, reflecting normal airport practice, had the passenger facilities, including duty-free sales, *after* the customer had passed through the British outward passport and customs checks. To facilitate the grouping of operations, it was agreed that the passenger facilities would be visited *before* the outward controls. This enabled the frontier and security controls to be located back to back and permitted greater flexibility in the design and organisation of the passenger facilities. An ancillary effect was that it was necessary to segregate traffic between that visiting the passenger facilities and that going direct to the shuttles as early as possible. This means that even before the tolls this separation has to be introduced. Management through the frontier/security controls is however much easier.

The second factor, shuttle design and loading regulation, had a much greater influence on the traffic management. Further, it took much longer to reach a design solution that was acceptable to the IGC.

Right from the initial proposal to governments the basic types of shuttle and railway carrier wagon have not changed. The three types of wagon are:

- freight, carrying HGVs up to 44 tonnes and 4.0m (later 4.2m) in height
- tourist single-deck, carrying passenger vehicles including coaches up to 4.2m high, caravans etc.
- tourist double-deck, carrying passenger vehicles on superposed decks and consequently limited in height to 1.85m.

These wagons were to be grouped into shuttle trains, made up of one or two rakes with each rake containing only one type of wagon. It was also agreed that freight and tourist rakes could not be mixed in a single shuttle. Since each rake is loaded from a separate point, the fundamental requirement was to segregate the traffic into vehicle types that could be carried in each type of wagon and this has remained unchanged. However, the following variations have taken place:

- freight shuttle, changed from a single rake of 25 wagons to two rakes of 14 wagons each, hence two loading points instead of one
- tourist rakes, changed from 13 short wagons to the current design of 12 wagons 26m long
- standard tourist shuttle formation of one single-deck rake and one double-deck rake, changed from having the double-deck rake at the front to the present arrangement of single-deck rake first and double-deck rake at the rear
- introduction of a loading wagon with an internal ramp permitting the development of a platform zone at a single level
- individual wagons closed off with a fire door at each end, with an additional fire door protection device before each barrier, reducing the space available for parking vehicles and increasing the need for accuracy in selecting the number of vehicles that travel in each tourist wagon
- single-deck vehicles loaded in such a way that any individual wagon would only contain either one large double-deck coach (61 seater or longer), one standard size coach and two normal cars, or two mini-coaches (16–40 seater).

The separation of vehicles arriving at the terminal into the three main types – destined for freight, tourist single-deck and tourist double-deck wagons – is achieved using conventional road signing. The height of the vehicles is checked at tolls, so that vehicles over 1.85m high are identified and informed at tolls enabling them to follow subsequent directional signing. In addition vehicles carrying liquid petroleum gas (LPG), essentially caravans and camper vans, have to be segregated.

The core of the traffic system is the subsequent separation and grouping of vehicles into a string of vehicles that could be loaded into the departing shuttle. The overall process is referred to as 'allocation'. In addition to the segregation of vehicles, the grouping process has to be synchronised with the timetable of shuttle departures. This 'timetable' is not published as is a normal railway timetable – the fundamental design is for a high frequency of departures permitting the simple turn-up-and-go travel pattern similar to a toll motorway – sometimes referred to as the 'rolling motorway'. Nevertheless the railway control centre must interleave the Eurotunnel shuttles with the high-speed Eurostar trains between London, Paris and Brussels, which in turn must fit into the timetables of their respective country train networks and schedules. Consequently the control centre operates to its own timetable for the Eurotunnel shuttles, which is transmitted via the RTM-TTM link to the terminal control centre (TCC; Figure 3) for the organisation of the traffic through the allocation process.

It was envisaged that vehicles would arrive at an allocation booth where the transport ticket issued at the tolls would be read and the individual vehicle directed to a specific slot in the loading pattern of a shuttle. This was subsequently changed so that vehicles would stream into the waiting area or reservoir by self-selection and following green light/arrow signals. Similarly, the possibility of measuring the length of each individual vehicle for double-deck wagons was investigated so that the number that would fit into each rail wagon could be calculated directly and the driver informed of the wagon he would be travelling in. It was, however, decided that the likely inaccuracy of any vehicle measurement system plus the uncertainty over the gap that could be achieved between vehicles when parked in the shuttles meant that such a calculation could never be sufficiently reliable. Hence it would be better to use the experience of the shuttle crew who were loading the vehicles to arrive at a typical figure for the total number of vehicles that could be carried and simply count these vehicles through the allocation barrier.

The vehicles assembled in the reservoir for a particular shuttle (Figure 4) have to be guided to the correct platforms for loading onto the shuttle. Since at peak times it is possible to have up to three shuttles (typically one freight shuttle and two passenger-vehicle shuttles) loading virtually simultaneously, it is the responsibility of the TTM to ensure that the reservoirs for the waiting vehicles are selected so that in driving to their platforms the vehicles are not forced to cross over other streams of traffic. The basic design of the platform

Figure 3. Terminal control centre, Calais. (Source: QA Photos)

zone in the two terminals facilitates this in that, although all tracks and platforms may be used by all types of shuttle and vehicles (giving considerable flexibility in operations and in planning maintenance etc.), the tracks normally used by freight and tourist shuttle are grouped separately. Thus, in the UK, freight tracks are always to the north of the tourist tracks. In France, where the terminals for processing freight and tourist traffic are completely separate, the freight tracks are always to the southeast of the platform zone. With the routes for vehicles established by the system, vehicles are released and drive to the platform either guided by a 'follow me' vehicle or following overhead gantry signs carrying the required platform number for each line of vehicles (Figure 5).

Operational system

The core of computer software relates to the allocation and loading procedures. It is linked to:

- the rail traffic management system (RTM) to receive details of the shuttle schedules
- the traffic control and supervision computer (TCSC) to receive statistical data on traffic arriving into the terminal
- the marshalling consoles (MC) which control the allocation of vehicles to shuttles
- the Eurotunnel MIS for reporting.

The TCSC receives via the RTM-TTM details of the next shuttles to arrive in the platform: type of shuttle, standard configuration or not, arrival time and

Figure 4. Reservoir lanes from the Folkestone Terminal control centre. Lanes are selected for vehicles to guide them smoothly to the correct platform for loading. (Source: QA Photos)

track. This is updated as the shuttle enters the Tunnel from the opposite terminal, i.e. roughly 30 min before arrival in the platform zone.

The TCSC must manage the traffic in the pre-ordering or marshalling areas. Vehicles segregate themselves into lanes according to vehicle type. The allocation process under the direction of the marshalling consoles releases the vehicles in the sequence required for loading into the shuttle. Vehicles are directed under the control of the TCSC into reservoir lanes where they will normally wait briefly until the shuttles are ready for loading. They are then released and directed to the correct one of the eight platforms from which to load into the designated shuttle.

While the marshalling consoles are responsible for the correct release of the vehicles from the pre-ordering lanes, the TCSC has to select the reservoir lanes for tourist shuttles: two lanes are required for each deck of the rake, hence six lanes for a complete shuttle. It also has to direct and synchronise the movement of vehicles from pre-ordering lanes to reservoirs and reservoirs to platforms so as to avoid crossing streams of traffic or vehicles descending the wrong ramp. The TCSC receives a shuttle-deck model giving the number of

Figure 5. Bridge to platforms with gantry signs showing platform numbers, Folkestone Terminal. (Source: QA Photos)

wagons in each rake and the number of vehicles that may be loaded into each wagon. The software contains algorithms to check the type of vehicle that may be loaded into each wagon. The deck model is passed to the marshalling consoles located in the tourist allocation (MC2, MC3) and freight allocation (MC4) booths.

The allocation can be performed in four different modes:

- Automatic: once activated by the controller in the TCC, the release of vehicles in the pre-ordering is performed automatically by the system.
- Supervised: the system prompts the marshalling console operator with vehicle release instructions, but the operator must confirm this promptly before the vehicles are released.
- Manual: the marshalling console operator has to be authorised by the controller for the mode to change between automatic and manual. Once in manual mode, the console operator actions the release of vehicles from the different pre-ordering or marshalling lanes in accordance with the wagon model on his console.
- Direct: in the event of failure of links or system, the system defaults to direct mode with the pre-ordering exit barriers raised and the barriers set at red. Vehicles must then be waved through by a traffic marshal.

At the start of the process of allocating vehicles to the next shuttle in the timetable, the TCSC assigns reservoir lanes for each deck. Vehicles are directed automatically, using green arrow routes on gantry signs in the UK and a system of lifting/rotating barriers and lights in France, into the correct lane. The system counts vehicles out of the pre-ordering lanes up to the limit of the first reservoir lane capacity and then changes the signs/barriers to the next lane. For tourist upper and lower decks, vehicles may be allocated simultaneously to upper and lower decks, necessary at peak traffic to achieve design throughput. At a set time before shuttle departure the TCSC prompts the allocation controller to release the vehicles from the reservoir lanes to the shuttle platforms. When the message is received from the platform (from either the train captain or platform supervisor) that the shuttle is ready to load, the controller confirms that a route has been correctly set between the reservoirs and platforms, and releases the vehicles, thus initiating the loading sequence (Figure 6). To reduce loading time the upper and lower deck streams of vehicles for the tourist double-deck shuttle are released simultaneously and drive along the overbridge side by side.

At this point the allocation to that shuttle would normally be stopped, although if there is no danger and no risk of a delay to the shuttle, the allocation controller can override the aborting of the allocation and continue to accept more vehicles, a practice only recommended in periods of low traffic.

Figure 6. Cars being loaded onto a double-deck passenger-vehicle shuttle wagon, Folkestone Terminal. (Source: QA Photos)

In addition to the allocation and loading procedures, the TCSC supervises:

- levels of parking occupancy compared to capacity to warn if use of emergency parking areas may be needed
- opening of lanes at the security and frontier controls area and checking for queues forming
- the unloading of vehicles from arriving shuttles (especially important on the French terminal where freight and tourist traffic leaves in opposite directions and variable message signs and barriers have to be set accordingly).

Finally, the TCSC centralises traffic counts from the tolls and from the shuttle allocation and stores them in a database.

SIGNS

Development

All aspects of signing were originally part of the terminal construction section of the contract. The design for the messages to be conveyed and the location of the sign structures was carried out under the direction of the TML terminal designers. However, as the award of the subcontracts on the two different terminals approached, the interactions with the TTM became increasingly apparent and the production and installation of traffic signs was transferred to the TTM contract.

Eurotunnel also had an interest in having this topic in the TTM contract since the traffic management principles were the same on both terminals and therefore required to be supported by a single coordinated signing approach that ran right through the system. It is likely that, if signing had remained part of the terminals contracts, the results would have relied heavily on the French and UK national road signing system. Eurotunnel wished to promote its own corporate image throughout the system, covering signs and graphics in the buildings and in the shuttles and felt strongly that road signs should form part of this corporate identity as in other transport systems such as Paris Métro or London Underground.

Eurotunnel therefore engaged consultants to produce overall corporate identity principles covering graphics, pictograms and colours and, through a

Figure 7. Signs in Eurotunnel design on approach to toll plaza, Calais Terminal. (Source: QA Photos)

variation order to the original TTM contract, obliged TML to incorporate these principles into the traffic signs (Figure 7). The initial work under the construction contract had concentrated on the message content at the various intersections and decision points through the road networks on the terminals. This was done in close cooperation between the two terminal teams so that commonality of directions could be achieved.

The detailed design work, still carried out by the separate UK and French teams, determined sizes and exact locations of signs. There are thus still significant differences in the appearance of the signs on the two terminals because the designers chose different sizing principles (e.g. according to perceived vehicle speeds). Certain general differences of philosophy were also apparent. Thus in France it was often not sufficient to indicate the travel direction (turn left); it was also necessary to have an additional sign prohibiting the alternative (no right turn).

Once the first stage of the detailed design was complete, the Eurotunnel signing concept, based on the corporate identity work, was applied resulting in some reworking of the detailed design to achieve similar presentations of general directional and informative signs.

The main link between the terminal design aspect of signing and the traffic management was through the variable message signs. These were required at the toll plaza to enable customers to select the appropriate vehicle route or method of payment, in the car parks to manage parking in traffic peaks, in the allocation area where the required lanes for traffic varies according to the shuttles to which they have been assigned, and on the routes from the reservoirs to the ramps down to the loading platforms. The control of these variable message signs, especially the last two, comes firmly under the traffic control system of TTM.

It was necessary to obtain the agreement of the respective governments to the use of the nonstandard road signs. It had at first been thought that the governments would not give their approval to any road signs that did not correspond to their respective national standards despite the fact that all the signs covered would be within the Eurotunnel Concession area and not on the public highway. However, following the submission of the integrated set of signs, Eurotunnel and TML were allowed to go ahead. The principles behind the final realisation of the traffic signs were set out in a signing concept submission by Eurotunnel to the IGC. The main features are:

- All statutory regulatory signs are implemented in accordance with national standards.
- General direction, information and vehicle selection signs are dark blue text on a white background.

- A specific family of pictograms was developed covering both vehicles and services.
- Wherever possible pictograms or a single word (e.g. services) understood in both languages would be used.
- For customers, if wording was necessary in both languages, it was presented first in Roman (upright) characters in the local language and in italics in the other language. In noncustomer areas only the local language was used.
- Univers typeface with lower-case lettering was used.
- A specific system pictogram was used to provide a general guidance symbol through the terminal to the shuttle.

Agreement was obtained to use this symbol on the national road networks to direct customers towards and into the two terminals. Although it was originally intended that 'Eurotunnel' would be accepted as the terminal destination on the national roads, agreement was finally given to 'Channel Tunnel' in the UK and 'Tunnel Sous La Manche' in France.

Operational installation

The principal means of guidance of vehicles through the terminals are the vehicle pictograms segregating freight and tourist vehicles, and differentiating vehicles for single-deck and double-deck shuttle trains, green arrow/red cross to select or prohibit routes through vehicle lanes (tolls, controls, reservoirs), and numerals to designate the platforms to which vehicles are directed for loading. The pictograms, although similar to those in general use in each country, are an original set of vehicle signs. They differ from the UK standards in that they are constrained to a square format and the vehicle silhouettes are more modern than the French standards.

There is some variation in the detailed design and application of road signage between the UK and French terminals, due to the different routes by which supply subcontracts were let. Those on the UK access road gantries are rather small, whereas those on the route to the French allocation area are more suited to motorway speeds than the 30–50km/h speeds that will be usual in that area. Similarly, for the green arrow/red cross signs, in some instances cold-cathode technology is used while in others signs are of the dot matrix type.

The numbering of the platforms and the dot matrix variable message signs guiding traffic to the platforms seems to work reasonably well. However, the means of passing this and other information to the customers while they wait in the allocation reservoirs is less satisfactory, being limited to a curt 'follow/ suivre' and a number. It is in this area also where it is most necessary to pass information to customers on the time when loading will start, the fact and cause of any delay, and instructions for loading. It is Eurotunnel's intention that the signing in this area will be made more extensive and informative.

In general, the intention was not to swamp the traveller with a forest of signs but to keep the direction to the minimum of vehicle type, green arrow, platform number. However, there are other needs for information to customers, for example: services available in the passenger terminal building, requirements of the French and British frontier control authorities, and exit routes for rejected vehicles. These may have made it more difficult for customers to follow the correct route but operational experience will doubtless show up those areas where improvements may be needed.

V
SYSTEM OPERATION

Operating the Tunnel

ALAIN BERTRAND AND ROGER HACQUART

There is no doubt that the creation of the Channel Tunnel, the two operational terminals, and the design and provision of the vast amount of equipment, both fixed and moving, is a tremendous achievement in engineering terms. However the ongoing task is to operate the system unfailingly and efficiently in order to provide the service which customers rightly expect, and to repay the investors and lenders for their confidence in the project from the outset.

There are two separate strands to the utilisation of the Tunnel, each of equal importance since they each have access to half of the capacity. The key strand belongs to Eurotunnel, since it is responsible for operating the Tunnel, the other jointly to the national railway companies of the UK, France and Belgium.

EUROTUNNEL'S OPERATION

Eurotunnel's railway, like any other, is set to provide a train service for its customers day and night, all year round, according to the timetable and automatic route-setting programs. However, there are special features that make operating the Channel Tunnel very different from running a conventional main-line railway.

First, there is close integration of fixed equipment with train movements. For example, in addition to signalling and traction power, the control of ventilation plays a vital part in the operation of a triple-bore tunnel 50km long. Air movement is controlled by the normal and supplementary ventilation fans at the two coastal shafts, by the operation of dampers in the 195 piston relief ducts between the two rail tunnels and the giant doors dividing the two crossover caverns and by the flow of air through 272 cross-passages from service to rail tunnel of which a proportion have controlled air bypasses, and finally by the speed and direction of the trains themselves.

In an emergency, the role played by all these components in ensuring that smoke is directed away from people is crucial. If workers have to enter a rail tunnel to rectify a fault, their safety depends not only on the diversion of trains to the other track, but also on wind speeds being controlled.

The controllers responsible are not just signalmen; they are qualified technicians who must understand precisely the way the trains and the fixed equipment they are controlling interact. Their task is more akin to the control of a large industrial plant such as an oil refinery.

Another factor is the extraordinary density of traffic. The design assumptions for fixed equipment were that the Tunnel would carry in each direction about 65,000 trains in 2003. This assumed 3 min headways yielding 20 standard paths an hour. Ultimate capacity, with headways reduced to 2 min giving 30 paths an hour, was put at 131,000 trains. A major upgrade of signalling and other fixed plant would be required to achieve this.

Admittedly, only shuttle trains occupy one path. Eurostar TGVs will go faster and through freight trains slower, so the number of trains passing in each hour is fewer than 20. Nevertheless, the Tunnel expects to be handling some 400 trains a day by 2003, in effect, running at metro frequency, but with a mix of train types that makes the whole operation far more complex; and, by European standards at least, they are large trains.

The planning of Eurotunnel's operations started from the realisation that very high standards of reliability were vital to the commercial success of the project. So far as equipment is concerned, Eurotunnel has invested heavily in redundancy to achieve reliability. As with a large ferry or plane, a shuttle train must work continuously through traffic peaks, receiving inspections and overhauls only when the demands of customers permit some respite.

The maintenance shops at the Calais Terminal, along with a light repair facility at Folkestone, are organised accordingly. Routine weekly servicing and replacement of defective modules is carried out without splitting the shuttle train, as experience has shown that it is best to leave couplings alone when possible to avoid introducing faults. Most fault finding and rectification will be carried out away from the train by swapping modules. Where the wagons have to be lifted for heavy maintenance in the Calais workshops, they can be jacked three at a time as triplet sets which again avoids needless disturbance to jumpers and wiring within the set.

Staff issues

The cost structure of Le Shuttle is different from a normal railway. Capital charges swallow the bulk of revenue, and energy costs are high: more than 20% of the total, compared to 6% on a typical railway. Therefore, it is worth investing heavily in staff of real quality to ensure that they respond in an intelligent and versatile way to minor problems as well as major crises.

Operations staff have been recruited into six departments. There is one for each terminal, handling the collection of revenue at the toll booths, the management of road traffic flows, and the marshalling of vehicles onto the platforms ready for loading. Four railway departments cover Le Shuttle crew, the railway control centres at each terminal, rolling stock maintenance and fixed equipment.

Train crew will number 600 as the planned service level for the early years is reached. This will see a maximum of eight passenger and seven freight shuttles in circulation at one time. Transit time between terminals is 35 min for both types. Layovers on the British side are about 20 min for passenger and 30 min for freight, more time being required for loading lorries correctly and transferring drivers to the club car at the front. Layovers in France are longer to provide a recovery margin. Hence, four passenger and three freight shuttles can be dispatched every hour, when demand warrants it. As demand builds up, more rolling stock will be ordered, and there is space at each terminal to double the number of loading platforms from eight to 16.

Each passenger shuttle has a minimum crew of eight: a driver in the front cab, a train captain in the rear cab, and six attendants spaced along the train who help with loading, and assist passengers during the journey while enforcing safety rules such as no smoking or working on car engines. Freight shuttles have a crew of three: driver, train captain and steward, with the last two riding in the club car.

Half of this workforce are trained as drivers, including all the train captains. At least two of the attendants on a passenger shuttle must be qualified as train captain, ready to take over that function if the appointed crew member has to drive in an emergency. A driver will typically spend 40% of his time driving, 40% as train captain and 20% as an attendant. This is to ensure that all crew members have a feel for the jobs their colleagues are doing, and also to combat the boredom of driving through a long featureless tunnel.

Versatility is again demanded in the technical staff. The most highly skilled are the troubleshooters stationed on the loading platforms to deal rapidly with any malfunction in the equipment on a shuttle. This demands an ability to cut across traditional craft demarcations and deal with electronic, electromechanical, hydraulic and pneumatic systems with equal facility.

Maintenance

Routine maintenance in the rail tunnels is carried out during timetabled 'possessions' every night. During this time, single-track working is in force between crossovers, allowing each of the six sections to be closed once a week in rotation.

Should it be necessary for staff to enter a rail tunnel at other times, that section is closed and a reduced service operated over the remaining track. In both cases, dampers in the piston relief ducts will be closed which leads to reduced train speeds so that workers in the other rail tunnel are not exposed to dangerous wind speeds.

Control centres

Smallest numerically of the operating departments is the control centre team, which is based in the control centre at Folkestone but also staffs the railway control centre at Calais.

Each control centre has three operations rooms:

- The **railway control centre** (RCC) on the ground floor can supervise train movements throughout Eurotunnel's territory and all fixed equipment in or affecting the tunnels.
- The **terminal control centre** on the top floor has a panoramic view of the road vehicle marshalling and queuing areas and is staffed by the terminal operations team.
- The **major incident coordination centre** is simply a large room with desks and communications which can be occupied during an emergency.

Folkestone is the 'master' RCC, normally responsible for operating the railway and supervising any maintenance of fixed plant in the Tunnel, except that Calais RCC supervises train movements within its own terminal. However, Calais is permanently on 'hot standby', ready to take over from Folkestone at a moment's notice if anything puts Folkestone out of action. Staff rotate between the two RCCs as part of their normal shift.

Up to eight people can work simultaneously in the Folkestone RCC; the minimum number on duty at one time is three. The 'slave' RCC at Calais is normally staffed by one person supervising the terminal area. However, there is an elaborate system for bringing additional staff into play quickly at either RCC at short notice if necessary.

Each RCC workstation has three screens, each connected to one of the three computer systems which control the central nervous system of the Tunnel: the railway traffic management system (RTM) which integrates timetable planning with direct control of trains through signalling, and normally works on the basis of automatic route-setting; the engineering management system (EMS), which controls all the fixed equipment except the signalling, including power supplies, ventilation, doors in the tunnels, pumps, cooling and fire protection; and communications display, which assists the controller to use the various telephone and radio networks available.

SERVICES VIA THE TUNNEL

Le Shuttle

The shuttle service, operated by Eurotunnel, accounts for half the Channel Tunnel's capacity. It is designed to transport passengers, cars, coaches, motorcycles and lorries/goods vehicles, on an on-demand basis. Journey time is 35 min.

In early 1995 passenger-vehicle shuttles are running approximately hourly during the day, less frequently at night. Freight shuttles are running half-hourly (for most of the day) or hourly.

Through trains

The nonshuttle half of the Tunnel's services comprises passenger and freight through trains connecting London, Paris, Brussels and later on other major destinations on the Continent. They are run by other rail operators, British Rail (BR, in the shape of European Passenger Services and Railfreight Distribution) and Société Nationale des Chemins de Fer (SNCF) most prominently, but also Belgian Railways (SNCB), Deutsche Bahn and Netherlands Railways. Eurostar day trains were introduced first, to be followed by overnight trains with sleeping cars. Through-service freight trains haul all sorts of goods (many containerised or swap-bodied) on any appropriate wagon. Extensive studies have been carried out to determine which goods, in spite of being classified as dangerous, can be accepted through the Fixed Link.

Eurostar services, via the Tunnel, link Waterloo International (and, from late 1995, cities further north like Birmingham, Manchester and Edinburgh) with Paris, Brussels and other Continental destinations. They are operated by BR, SNCF and SNCB. Trains join the new TGV Nord Europe route at Fréthun, near the Calais Terminal; the new route runs via Lille to the Gare du Nord, Paris. Trains for Brussels diverge from the Paris line near Lille. It is also possible to make connections with other TGV services at Lille and travel to other destinations in France, bypassing Paris. At their fastest, Eurostar trains take just under 3h to reach Paris (and 3h 15 min to Brussels), from London. However, improvements to these times are expected in the near future: the Belgian high-speed line (due in service in 1996-97) will cut the journey to Brussels to 2h 40 min. In the UK, eventual construction of the much-debated high-speed link between London and Folkestone, which will enable Eurostar trains to give their best performance, will reduce journey times by 33 min. From eight trains per day in late 1994, Eurostar services will be building up to a level of 30 trains per day.

Overnight services (with sleeper and couchette facilities) will start from three regional centres in the UK (Glasgow, Plymouth, Swansea), stopping at other important stations *en route* to Waterloo, from where they will travel to Paris and Brussels. There will also be overnight trains from Waterloo to Amsterdam, Dortmund and Frankfurt, and to Cologne. Corresponding services will run in the opposite direction.

The pattern of rail freight transportation between the UK and the Continent should be transformed by the Tunnel. The main complication is that all wagons need to meet the smaller British loading gauge (most Continental wagons are to UIC standards). Special 'Multifret' wagons have been ordered to cope with this requirement. Transfers of cars from the factory for distribution will be a major activity via the Tunnel: purpose-built car carriers have been commissioned for this purpose. In the UK, Continental freight services will operate from seven terminals around the country (two more planned). On the Continent, through freight train destinations will be determined by demand. Very substantial savings in time can be made by transporting freight via the Tunnel and the network of intermodal services: for example, Birmingham to Vienna (66h by road and Channel ferry, down to 31h) and Glasgow to Milan (72h to 40h).

THE NATIONAL RAILWAYS

When efforts were being made in the 1960s to create a Channel Tunnel in the form seen today, the primary source of revenue was considered to be the carriage of road vehicles on shuttle trains. At that time, rail was widely viewed as an obsolescent mode of transport, although its share of the total passenger and freight markets was substantially larger than it is today. By the early 1970s, however, the concept of building new high-speed railways able to compete with short-haul airlines was gaining ground in Europe. One such line was proposed between Paris and London in 1973 but the project had collapsed by 1975.

Two national railways, BR and SNCF, therefore adopted a modest and cautious approach when they put forward in April 1981 plans for a single-track Channel Tunnel with no road shuttle service. This would have carried high-speed trains on a new TGV Nord line from Paris to Calais, but they would have reached London by using the existing lines electrified at 750V DC on the third-rail system. The Tunnel, like the connecting link to Paris, would be equipped with overhead catenary energised at 25kV AC. In addition, a new international station was to be built somewhere near the junction between the cross-channel link and the existing Ashford–Folkestone line. Apart from providing interchange with local rail services, as Fréthun does today on the

French side, this station was to be the starting point for long-distance sleeper and motorail services using standard Continental rolling stock.

BR's bridges and tunnels are generally lower and narrower than the most restrictive Continental gauge, but rail freight already using a train ferry link was to be expanded using wagons of British gauge. There would be more emphasis on intermodal and block trains.

When the Channel Tunnel Group/France Manche (soon to become Eurotunnel) proposal was chosen by the two governments in January 1986, the international station had been designated as Ashford and was no longer intended to act as a starting point for sleeper or motorail services. This meant that Continental-gauge rolling stock now had nowhere to go if it came through from France, as there were no plans to handle rail freight or passengers within Eurotunnel's Folkestone shuttle terminal.

So the entire Continental rolling stock fleet was effectively barred from the Tunnel, with the exception of some wagons that used the Dover–Dunkerque train ferry. How was it that the long-planned dream of uniting the two rail networks failed to come about in the way that had been envisaged?

Safety issues

It was not just the tight loading gauge on the British side that barred Continental carriages from the Tunnel. As it turned out, safety provisions imposed on the operators of through passenger and freight trains meant that no existing coaches or locomotives, British or French, could be used. Thus, to initiate the service, nine BB22200 class SNCF locomotives were adapted to haul freight trains until the 46 purpose-built Class 92 locomotives became available, but this was only a temporary arrangement.

Conditions with which passenger and freight rolling stock used in international service must comply are laid down in Europe by the International Union of Railways (UIC). Various international agreements cover, for example, the carriage of dangerous goods by rail. In principle, rolling stock and freight which meet these requirements can go anywhere, even through Alpine tunnels up to 20km long. Likewise, the 54km Seikan undersea tunnel which links the islands of Honshu and Hokkaido in Japan is open to normal passenger and freight rolling stock.

However, the Channel Tunnel is unique. The special safety rules imposed in it stem from the sheer scale of operations. A fully loaded shuttle carrying 120 cars and 12 coaches could have 1000 people on board, while a Eurostar speeding from London to Paris can carry 800. The Channel Tunnel may not be the world's longest rail tunnel, but it does feature the longest distance between ventilation shafts (39km), and it is destined to carry unprecedented levels of traffic in years to come.

Special rolling stock

Leading the fleet of purpose-built rolling stock which is permitted to pass through the Tunnel are the Eurostar trains (see also Chapter 12). The French, British and Belgian railways have ordered a total of 38 trains. Based on the French TGV, the Eurostar can run at 300km/h but it will be restricted to 160km/h in the Tunnel, mainly for 'pathing' reasons, and when drawing 750V DC power in Britain. The Tunnel is aligned for 200km/h operation should this be considered practical in the future. These intercontinental trains run in France on the dedicated high-speed line.

In addition to the Eurostar, a second fleet of passenger rolling stock, which will provide overnight services from 1996, has been ordered. Although BR had a surplus of fire-hardened sleeping cars introduced in 1982 (after the fatal Taunton fire of 1978) these were not deemed to be safe enough to pass through the Tunnel. Hence an order was placed in July 1992 for 139 sleeping, sitting and service cars to run overnight between British and Continental cities. Like the Eurostar trainsets, these will be formed into two half-trains of seven or eight cars with automatic uncoupling from inside the train. However, they will be hauled by a variety of locomotives, including Class 37 diesels to Swansea and Plymouth, so unlike Eurostar they have to conform to UIC specifications for coaches circulating on the Continent.

When passing through the Tunnel, each overnight train will have a Class 92 electric locomotive at the front and rear. With automatic uncoupling of these locomotives available, a similar configuration to the Eurostar sets will be achieved: i.e. either locomotive can be shed if it catches fire or becomes incapable of being moved, or the train can be split in the middle after passengers have been evacuated from one half into the other. Class 92 is a dual-voltage Co-Co locomotive ordered originally by BR to handle international freight services within Britain and through the Tunnel to Fréthun (see also Chapter 12). A total of 46 identical units has been ordered, nine of which will be owned by SNCF and seven by European Passenger Services for the overnight trains.

While all passenger carrying trains must have a locomotive or power car at each end, freight trains have only the one driver on board. The temporary BB22200 locomotives in initial service will always be used in pairs, but Class 92 may be used singly or in pairs depending on the load. To enhance reliability when used as a single locomotive, the Class 92 body has internal bulkheads capable of 30 min fire resistance.

COMBINED OPERATIONS

Timetabling

While Eurotunnel is offering an on-demand service to road users, Le Shuttle operates to a timetable planned many months in advance. The process starts with agreement at international conferences in September each year on the timetable for through passenger trains that will run for 12 months from the following May. Draft timetables and rosters for crew and trains, with loading platform occupation specified, are circulated before the final base timetable is agreed 30 days ahead of service. This is used by the RTM for automatic route setting, but it can of course be edited as required up to and including real time changes necessary to cope with delays to trains as they occur. To assist the controllers in regulating the flow of trains, routines have been prepared which automatically adjust schedules to take account of deviations from the timetable.

National railways

Trains passing through the Tunnel fall into one of four speed bands. Eurostar runs at 160km/h and takes 18–20 min between portals. Class 92 is limited to 140km/h and will take overnight trains through in 22 min. Freight trains hauled by Class 92 are limited to 120km/h or 100km/h, according to the wagons being hauled. A train weighing 1600 tonnes gross will take about 32 min.

Eurotunnel's timetable is based on shuttle paths, which assume that a train running at a maximum speed of 140km/h will transit the Tunnel in 26 min. Eurostar and freight trains can therefore occupy more than one shuttle path. For this reason, Eurostar trains are scheduled to pass through in pairs, 3 min apart, bound for or coming from Paris and Brussels, respectively.

Signalling

Like TGV Nord, the Tunnel uses TVM430 in-cab signalling so there is little interface to be handled on the French side as Eurostar trains enter or exit the Tunnel. All other trains stop at SNCF's Fréthun yard for locomotive change.

Class 92 locomotives, like the Eurostar power cars, are equipped to receive and interpret TVM430 track circuit codes. In the case of the locomotives, however, the driver must enter a description of his train (maximum speed, passenger or freight) before leaving the yard.

The position at Folkestone is more complicated. As regards traction, the Dollands Moor sidings are equipped with both 25kV catenary and 750V third rail. Eurostar trains change traction voltage on the Continental main line at 160km/h, there being an overlap between the two systems. Commands conveyed by the inductive cable loops forming part of the TVM430 signalling drop the pantograph and lower the collector shoes of a Eurostar train heading towards London before it runs off the wires; a similar traction changeover is initiated in the opposite direction.

One potential difficulty for controllers is the very short distance between the point at which through trains arrive on Eurotunnel territory and the point at which tracks converge at the tunnel portal. On the British side, Ashford signal box controls the principal route to Folkestone, including the departure signals for freight trains leaving Dollands Moor yard for France. On the French side, it is the new signal box at Fréthun which hands over both freight and TGVs to Eurotunnel, even though most of the new TGV route is controlled from Lille. Thus there is a double transfer of responsibility that requires close co-ordination.

TVM430 loops are also used to 'arm' and 'disarm' the cab signalling equipment on trains as they traverse the transition between BR's automatic warning system and Eurotunnel's tracks, or leave Fréthun yard to enter the Tunnel.

22

Safety management

RICHARD MORRIS

Although intrinsic safety was an important factor favouring the two governments' choice of a twin-bore rail tunnel for the fixed link under the Channel, the issue of safety management was first officially addressed with the signing, early in 1986, of the Treaty of Canterbury and the Concession Agreement which established an Intergovernmental Commission (IGC) charged with accepting the detailed design of the Project. The IGC could withhold acceptance of any aspect of Eurotunnel's proposals on safety, security, defence or environmental grounds, but it was the safety issue that proved to be the biggest and the most expensive.

The Safety Authority (SA), which advises the IGC, had to be convinced that any risk to passengers and staff had been reduced to an acceptable level. However, the Concession Agreement had implied that control by the IGC and SA would be light: for example, failure to object within 15 days to a particular outline design signified acceptance. This rule was quickly abandoned in favour of a rigorous and often protracted approval process.

What has made the Channel Tunnel project so unusual is the fact that the broad safety standards to be applied were not established in advance of the Concession Agreement. The result was a moving target that became ever more demanding and costly to achieve as design and construction progressed.

The SA sought constantly to eliminate any risk or combination of risks which could be identified. While the financial consequences for Eurotunnel were serious, leading the two governments in December 1993 to extend the Concession by ten years by way of partial compensation, one positive result has been to create a culture within the organisation that promotes safety at every level.

Moreover, quite apart from the legal and moral obligation laid upon any transport operator to strive for high levels of safety, in the case of Eurotunnel they are vital to commercial success. There is a popular perception that a long undersea tunnel must be inherently dangerous, and experience during the construction phase and at the start of operations demonstrated that media coverage of even minor accidents is intense. Because of the fiercely competitive market, such publicity could have a negative impact on revenue out of all proportion to the real threat to life.

The reality is that tunnels offer a protected environment where many common factors in railway accidents, such as level crossings, vandalism and weather, are eliminated. The best evidence that carrying people through long tunnels in trains equipped with automatic train protection (ATP) presents extremely low risks is provided by Japan's shinkansen ('bullet train'), since almost a third of the 1835km network is in tunnels up to 22km long. In 30 years, some 1000 billion passenger-km have been achieved at more than 200km/h without a single passenger fatality. This volume of travel, if undertaken by road in cars, would typically result in 10,000 deaths.

A specific danger in tunnels arises from smoke or the release of dangerous gases. Although hazardous freight is banned from the Channel Tunnel for this reason, it could be argued that intensive shuttle services for cars, buses and lorries pose a much greater hazard than shinkansen trains. However, the safety record for shuttle operations is also good. Carrying cars, buses and lorries through Alpine tunnels 15–20km long has been big business since the 1960s, with some 30 million vehicles transported to date. The largest operation is through the 15km Lötschberg tunnel, where shuttle trains carry around a million vehicles a year, mostly cars which park on open-sided wagons with the occupants sitting inside. There has been no accident to a train carrying road vehicles inside any Alpine tunnel that resulted in casualties, although there was a collision at a terminal station in 1992 in which a passenger died. This would not have occurred if ATP had been installed, as is the case throughout the Channel Tunnel and Le Shuttle terminals.

To ensure that the Channel Tunnel not only meets stringent risk criteria demanded by the Safety Authority, but also inspires confidence among potential customers, six major factors have been brought into play:

- a relentless drive to design, engineer and manufacture safe equipment and provide a safe system by planning out hazards
- recruitment of high-calibre staff with the right personality profile whose training includes a proactive safety culture
- production of the first full Safety Case ever to be devised and applied in the transport field
- preparation, by the future operators, of the rules governing Eurotunnel's operations for approval by independent safety experts
- a dynamic safety management system which embraces a quantitative audit process to measure safety achievement
- accident/incident analysis designed to discover the causes of any problems.

The last point is very important. It has been estimated that hazardous incidents outnumber accidents by 600 to 1, so reporting and investigating incidents yields 600 times more statistical data than rare accidents. To obtain this information, a caring and trusting culture is required. The 'blame culture' traditionally found on railways tempts staff to conceal incidents for fear of being disciplined. Eurotunnel believes that, if a genuine mistake has been made, and it is properly and honestly reported, no blame should be attached. However, a strong differentiation must be made between genuine mistakes and misconduct involving wilful disregard of safety rules or procedures, where disciplinary action is essential.

TRIPLE-BORE ADVANTAGE

Ever since work on a Channel Tunnel first began in the 1880s, most competent proposals have been for twin-bored rail tunnels rather than a single bore containing two tracks. The principal reason was geological: the desire of engineers to stay within a particular band of impermeable chalk marl favoured bores of small diameter. The arrangement finally implemented, two rail tunnels combined with a central service tunnel, linked to both rail tunnels at 375m intervals has significant safety advantages. The risk of a derailed train being struck by one coming the other way is obviated. Access by maintenance staff to sections of tunnel carrying trains can be banned as possessions can be planned on a nightly basis with two-way working through the other bore.

Ventilation of all three bores can be achieved by constantly feeding fresh air into the service tunnel. By controlling the release of air into the rail tunnels, the service tunnel remains pressurised, and thus forms a safe haven from a train on fire in a rail tunnel. Smoke in one rail tunnel can be kept out of the other by closing dampers in the piston relief ducts joining the rail tunnels at 250m intervals. Finally, emergency services have ready access to any incident through the service tunnel, even if both rail tunnels are occupied by trains.

Two sets of crossovers split the running tunnels into six segments. To maintain separation for aerodynamic purposes, the two crossover caverns are divided longitudinally by huge pairs of sliding doors (Figure 1). Interlocked with the signalling, they are opened only to permit trains to switch from one rail tunnel to the other.

Even earthquake risk has been assessed, although the last significant shock recorded in this area was in 1531 and considerable effort has been expended to estimate its magnitude. Though tunnels themselves are resistant to earthquakes, steps were taken to minimise possible damage to equipment installed in them, and to ensure that entrances could not be blocked by landslips or local collapse.

Figure 1. Pairs of 60-tonne longitudinal crossover doors which normally separate air flows in the two rail tunnels are opened to permit trains to switch from one rail tunnel to another. (Source: QA Photos)

FIXED EQUIPMENT

Up to now, the maximum load carried by a single track has been around 100 million gross tonnes a year on a heavy-haul railway in the USA. Eurotunnel expects to attain that figure within a decade or so, but running at much higher speeds. Ultimate capacity of the Tunnel is considered to be some 240 million gross tonnes a year over each track. With such dense traffic in prospect, safety and reliability demand track laid to exceptionally high standards (see Chapter 9). Rails of 60kg/m are secured to precast concrete blocks surrounded by a rubber boot, and cast into a solid concrete bed. The objective is to ensure that the rails cannot deviate from the correct line and level, yet there is resilience to temper dynamic forces which could fatigue and crack the rail. Should a derailment occur, despite hotbox and dragging equipment detectors, concrete walkways on both sides of the rail tunnel will keep vehicles upright and in line.

Eurotunnel's appetite for electric power is prodigious. The ultimate capacity is 360MVA for traction and 60MVA for pumps, fans, lighting etc. Power is taken direct from the national grids on each side of the Channel, and it is possible to run a reduced train service using power drawn from one end only. All other plant is fed by cables distributed between the three tunnels. At each of the coastal shafts, 5MVA of diesel generators can be connected to these cables to maintain essential services such as lighting, pumping and signalling in the highly improbable circumstance that both grid supplies fail.

ATP is provided throughout Eurotunnel's rail network by TVM430 signalling that was developed for the new TGV Nord line from Paris to Calais (see also Chapter 10). This uses coded track circuits to transmit data to computers on the train, which then calculate braking curves necessary to comply with speed restrictions and avoid collisions.

Trains running at the same maximum speed can be scheduled at 3 min intervals. If a driver fails to keep the speed below the limit indicated on the display, braking will be applied automatically. However, the ability to drive at 35km/h even when receiving no code is retained so that trains can still keep moving if there is a failure of the signalling system. This is known as *marche à vue* (driving on sight). The risk of collision under the ATP regime is very small. The risk of a low-speed collision during *marche à vue* is obviously greater, but serious consequences are only likely to occur if such a low-speed collision results in a fire, which is unlikely.

There is a control centre at each terminal, each with three operating rooms. One room supervises road traffic flows and other local functions at the terminal, but not the movement of trains. The second supervises the railway, power supplies, and all plant in the tunnels, with Folkestone normally in charge and Calais on standby. The third is a major incident room which is available for use by Eurotunnel personnel and local emergency services should the need arise.

COMMUNICATIONS

All plant, the railway signalling and an extensive telecommunications network is digitally controlled using three fibre-optic cables which form the spinal cord of Eurotunnel's central nervous system. They are contained in steel pipes, one in each tunnel, for maximum protection from physical damage. Optical fibres are immune from electrical interference, and if a cable is cut there is no 'short circuit' so signals can still reach receivers on each side of the break from opposite directions. A high level of redundancy is built in, such that the whole system could continue to function even if two of the three cables were each cut in one place.

There are numerous telephones throughout the terminals and the tunnels, interconnected to the national networks on both sides of the Channel. In addition, multiple independent radio systems are available (see Chapter 8). The objective is to ensure that communication between staff and the control centre is maintained at all times.

FIRE PRECAUTIONS

Apart from trains and their contents, electrical equipment is the most likely source of a fire in the tunnels. Technical equipment rooms, mostly excavated between the service tunnel and a rail tunnel, are fitted with optical and gas ionisation fire detectors. These activate a fire suppressant as well as sounding alarms locally and at the control centre. On the passenger shuttles and through trains, fire is likely to be detected by those on board. This is not necessarily the case with freight trains and freight shuttles so smoke detectors have been located just inside cross-passages, at 1.7km intervals throughout the rail tunnels. Normally, if an alarm is received the suspect train would be told to keep going at reduced speed, while those behind it would be stopped. Piston relief duct dampers would be closed to stop the smoke affecting the opposite running tunnel. Should it be necessary to fight a fire in one of the tunnels, a 250mm diameter water main feeds hydrants spaced at 125m intervals in all the rail tunnels.

The traditional approach to railway safety has been to stop a train if any potentially hazardous failure occurs. For obvious reasons, a quite different philosophy developed in aviation, where safety is achieved by redundancy coupled with isolation of damaged equipment so that it does not degrade healthy systems. Eurotunnel's approach falls somewhere in between these two.

Channel Tunnel motive power and rolling stock is unique in that it is designed to keep running for up to 30 min after a fire has been detected. The first objective is to get the suspect train out of the tunnel and into a special siding at the terminal where the emergency can be dealt with in the open air. Meanwhile, passengers are moved away from the fire to other parts of the train. But if this is impossible, a stalled train can be split, allowing passengers to escape in the mobile portion or, if that fails, they can be evacuated into the service tunnel.

This has had a major impact on the cost and complexity of rolling stock, most notably on the passenger shuttle wagons where the presence of petrol in cars poses particular problems. Fire barriers are provided at the end of each passenger wagon or coach so that the adjacent vehicle becomes a safe haven. The roof and floor must also be able to resist fire penetration for 30 min, so that a fire above does not affect brakes or cables below or, conversely, an underfloor fire does not penetrate the passenger compartment.

All materials used are fire resistant and produce minimal smoke when heated. The entry of air from the tunnel must be controlled so that vents can be closed to prevent smoke from another train or vehicle coming in. It must be possible to uncouple sections of the train from inside so that burning vehicles can be abandoned.

See also Chapter 16 dealing specifically with fire detection and suppression.

Passenger-vehicle shuttles

Aboard the passenger-vehicle shuttles, the first line of defence is prevention. Smoking is banned, and motorists are forbidden to lift the bonnet of their vehicles so as to reduce the risk of petrol or electrical fires being initiated. Six attendants spread along the train, backed by closed-circuit television, will ensure compliance.

The second line is detection. Hydrocarbon vapours, smoke or flame will automatically trigger a warning to passengers to evacuate the wagon immediately; tests have shown that evacuation takes 2 min for a double-deck wagon and 3 min for a single-deck wagon containing a coach. Although fire-resistant shutters will have closed off both ends of each wagon before departure from the terminal, passengers can escape through pass doors on each side.

The third line is suppression. Cars and coaches sit in a shallow trough which drains any leaking fuel into a holding tank. Should this fuel ignite, acqueous foam is injected to smother the flames. If a fire became life-threatening, a discharge of Halon 1301 would occur automatically. Though this may be

Figure 2. Passenger shuttle wagons have 30 minute fire barriers at each end with 700mm-wide pass doors. (Source: QA Photos)

alarming for passengers, tests have proved that, in the likely event of discharge over passengers no lasting harm will result.

The fourth line is containment. The wagon body with its ends closed by fire barriers should contain a burning road vehicle for 30 min giving the train time to reach the emergency siding in the terminal (see Figure 2).

The fifth line of defence is evacuation of passengers to another part of the train, which is then uncoupled from the burning vehicles, or into the safe haven of the service tunnel if the complete train is immobilised.

Freight shuttles

At one time it had been the intention to enclose the freight shuttle wagons so that Halon 1301 could be released if fire was detected, even though lorry drivers would be riding in a club car at the front of the train. In view of the trouble-free record of piggyback services in Europe, a decision was taken in 1989 to use open-sided wagons without fire suppression on board. If a lorry were to catch fire, the smoke would be detected by sensors mounted on each of the four shuttle loading wagons, and by sensors mounted within the tunnel. Depending on the number of sensors triggered, the control centre will initiate one of two alternative responses:

- **Status 1:** Trains following the suspect train are instructed to stop. All other trains in both directions reduce speed to 100km/h so pressure relief dampers can be closed, limiting smoke to one tunnel. All trains close their air intakes. The suspect train is routed into the emergency siding at the terminal.
- **Status 2:** Evacuation of the tunnels is ordered. The suspect train stops with the club car opposite a cross-passage. Trains trapped behind it reverse out. The front locomotive and club car are uncoupled and then exit the tunnel, leaving any fire to be extinguished from the service tunnel. If uncoupling fails, all personnel on the train evacuate into the service tunnel.

Through trains

Railway carriages carrying passengers through the Tunnel have been purpose built to the highest fire standards, as on a metro. Automatic extinguishers are confined to the power cars where the greatest risk occurs. Each carriage has a 30 min fire barrier at each end so that passengers can be evacuated into the adjacent coaches. The floor will also resist penetration for this time.

With important categories of hazardous goods banned, the risk to life presented by freight trains is minimal. Statistics suggest that the vehicle most likely to catch fire is the locomotive, but such fires do not as a rule develop rapidly and the driver can evacuate into the service tunnel if necessary.

Service tunnel transport

While service tunnel vehicles are normally to be used to take maintenance crews to their work, they provide the principal means of access for firefighters, medical teams and others who need to be on the spot quickly in an emergency. Travelling at 80km/h, they can reach the midpoint of the service tunnel in just over 20 minutes from passing through the airlock at the portal.

The Swiss and Germans have fully equipped rescue trains stationed near long tunnels, but many of these are double-track bores. The service tunnel transport system (STTS) was considered to be more flexible in that the vehicles can pass each other and reach the cross-passage where they are needed without being blocked by other trains. For evacuation of passengers from the service tunnel, by far the most effective method is to send in an empty shuttle train through the opposite rail tunnel. The STTS will not normally be used for this purpose, other than to take those injured to hospital.

HAZARD PRECAUTIONS

Eurotunnel has banned a wide range of hazardous commodities from Le Shuttle and through freight trains. Generally, these are products which would either burn fiercely if ignited, or they would render the atmosphere inside the Tunnel toxic if released. The ban extends to nuclear flasks, even though the probability of release is considered by the authorities to be acceptable for transit by rail through urban areas.

Open-topped wagons carrying loose powdered commodities may not be accepted without some sort of cover. This is because the jets of air from the piston relief ducts could blow material out of the wagon, creating dirt in the Tunnel which might interfere with the large amount of equipment installed in it. Live animals are banned, essentially for humanitarian reasons, although in practice this restriction only affects Le Shuttle.

There was concern about the effect of the piston relief duct air jets on curtain-sided containers and swapbodies. A design study showed a need for air flow restrictors, and these were installed. Tests confirmed that they reduced air flows to levels that were acceptable.

New cars can have a small quantity of petrol in the tank. Containers or swapbodies can pass through with diesel powered refrigeration units in operation. The freight shuttles have a train-lined power supply and plug-in points for refrigeration units on lorries and semitrailers.

EVACUATION STRATEGY

With fire present in a tunnel extending 39km between shafts to the open air, the need to move people away from flames, smoke and toxic fumes as quickly as possible is self-evident. This is why staged evacuation of passengers and staff forms a key element in the safety strategy developed by Eurotunnel. The principal stages are:

1. Evacuation from a burning vehicle into an adjacent vehicle while the train continues to the terminal.
2. Evacuation from the burning section of a train into another section, and abandoning the burning portion in the tunnel.
3. Evacuating people from an immobilised train into the service tunnel, which is kept under positive air pressure.
4. Clearing all other trains out of the tunnels, by reversal where this is necessary.

Stage 2 demands the ability to uncouple from inside the train in case the atmosphere outside is hostile. Passenger shuttles are formed into three-wagon sets for this purpose. Coaches in the Eurostar and overnight passenger trains will be formed in two sections, with the locomotives or power cars at the outer ends also capable of being detached from inside the train.

Stage 3 should be a very rare event. Where possible, passengers will pass through the train to alight on to the walkway close to a cross-passage into the service tunnel. Directions will be given by loudspeakers on the train, and from the control room using loudspeakers inside the service tunnel. Once in the service tunnel, passengers should be safe from harm. Normally they will be evacuated by a rescue train sent into the other rail tunnel, but rescue vehicles can be sent into the service tunnel if required.

To cater for Stage 4, passenger trains must have a locomotive and somebody competent to take the controls at both ends. Tracks are fully signalled for running in both directions, so there is no difficulty about getting the rear portion of a disabled train, and all trains stopped behind it, back to the terminal from which they entered the Tunnel.

The rear locomotive on freight shuttles is not manned. Freight trains will normally have one or two locomotives at the front end, and only a driver on board. In both cases, drivers will be instructed by radio to drive out in reverse when necessary.

The catenary supplying power for traction to the trains is split into sections 1200m long so that if it is damaged, generally only one train will lose power. However, the evacuation strategy includes the use of Eurotunnel's five diesel locomotives to haul trains out where electric traction is not available.

EMERGENCY SERVICES

There are well-tried procedures within the UK and France for dealing with major emergencies, but they do differ in certain key respects such as the chain of command when more than one service is involved. Some adaptation of these procedures has been necessary in the form of a binational emergency plan.

To test this plan, binational emergencies are simulated periodically in the form of command procedure exercises, an innovative technique which won Eurotunnel a national training award in 1993. This recognition is just one of many reasons why the Channel Tunnel and Le Shuttle is not only the safest railway system, but also the safest transport system ever conceived.

23

Conclusion

COLIN J KIRKLAND

The successful completion of the Channel Tunnel will always represent, for those whose efforts brought it about, a high point in their careers and an almost incredible engineering achievement.

In any league table of large projects, created for the use and service of man, it must rank with the highest. And yet, through most of the comparatively short period during which it was constructed, it seemed to be constantly under attack from the media – bedevilled by delays in construction, spiralling costs and shortage of funds. That this should have been so is partly due to the normal propensity of the media to see bad news as the best news, but it also has a lot to do with the tremendous risks that all parties took at the outset, in their enthusiasm to see the Tunnel built. We had waited 200 years for the political and financial climate to come right, and the opportunity was not one to be missed!

In the short space of time – six months – during which the bid proposals had to be prepared, fixed times and prices had to be assigned to concepts rather than to carefully specified tasks. Every aspect was pared down in order to be able to demonstrate the feasibility of such a huge undertaking, to be built entirely without public financial support.

Looking back, the construction phase was not a series of disasters, but rather a series of triumphs over adversity.

The Tunnel – or rather the three tunnels – required no fewer than 11 hybrid boring machines to be built and put into service, and two huge factories to be commissioned to provide the Tunnel's lining. Small wonder that, after 12 months of fevered activity, the work was eight months behind schedule. What is truly amazing, and almost always forgotten, is that, in spite of these early problems which are not abnormal on major tunnel projects, the Tunnel itself was completed some five weeks ahead of the original programme.

The construction of the two road/rail terminals at either end of the tunnel, each of the size and transport capacity of a major international airport, was completed within a year of their original seven-year programme. This despite having to contend with a multitude of modifications, from major traffic layout changes, and complete reversal of site drainage arrangements at Folkestone, to wind protection fences on the exposed French terminal.

Final commissioning delays are largely attributable to the complex nature of the rolling stock, which is unique worldwide, and an absolute commitment to public safety.

Turning to the subject of cost increases, again the public image of the project suffered through comparisons made in the media of sums that were not capable of comparison. For example, the contractors original estimate of construction cost – £2.8 billion – has been compared with the final total funding requirement of £10.5 billion. In fact the comparable sums for construction are £3.8 billion at the outset, including a provision for inflation, and £5.8 billion on completion. For financing we should compare the £10.5 billion final value with an initial funding requirement of £6 billion.

Since this book is about engineering the Channel Tunnel, detailed analysis of the reasons for these increases will have to wait for another volume. However, the main reasons lie firstly in the complexity of tunnelling operations where the cost risk was shared between client and contractor. Secondly, in the cost of making changes to the original scope and detail of works on the two terminals. Thirdly, in the cost of the rolling stock, for which a provision was made at the outset which proved to be too low by a factor of three when the detailed requirements became known as system design developed. Finally, the cost of financing rose significantly due to increased need for funds to cope with construction cost increases and delay in commissioning.

The pressure on all concerned to meet delivery deadlines and control spend within budget was immense, both upon the companies concerned, and perhaps more importantly upon individuals. Many who joined the team were forced to leave by the unrelenting pressure or the major management changes which had to be made to improve overall performance.

Evidence of these pressures may be detected in some of the chapters contributed. Many felt frustrated in their efforts to prepare designs following their normal logical sequence, by the continual changes in requirements brought about by work going on elsewhere on the project. Many design procedures which we would call normal in the leisurely development of, for example, a major highway project, had become luxuries that we could not afford. Some found the pace, and contending with constant uncertainty, exciting and challenging. For some it was just too much.

The question is often raised: 'If you could do it all again, would you do it differently?' The answer is certainly 'yes' in some respects, though, because we do not live in a perfect world and have to make the best of situations as they are, not much would change.

If we had not put ourselves under the pressure of cost and time, the project would never have been completed in such a short period, but much of the frustration brought about by change might have been avoided if the project had been predicated on the requirements of a transportation business rather than on the civil engineer's definition of the works.

A final thought for the future may be prompted by the little-known requirement upon Eurotunnel in the Concessions to develop proposals by the year 2000 for a 'drive-across' link. There are two key questions to be answered. Firstly, will a drive-through scheme prove feasible, and will it be a tunnel or a bridge? There is little doubt that a Tunnel of appropriate proportions could be driven, since the first Tunnel proved the geological profile and practicability. However, ventilation problems will remain insuperable just as long as our vehicles are powered by internal combustion engines, producing noxious gases. A bridge across the Channel is technically feasible, but the navigation hazards alone may well rule it out.

The second question is one of user safety. Under the Eurotunnel system there may be up to 20,000 people in the Tunnel, under the very carefully monitored control of about 15 'drivers', with a vast array of devices to prevent collision and detect and deal with fire. In a drive-through scheme there could be up to 12,000 drivers, each responsible only for themselves and a few passengers, and subject only to advisory signals. Furthermore, recovery of breakdowns and the traffic problems associated with even minor 'shunts' would result in totally unacceptable safety hazards.

However, while a drive-through scheme may not be something for us to look forward to, it is highly probable that a second pair of railway tunnels will have to be driven, perhaps within 25 years of the inauguration of the Eurotunnel service.

We should not leave this volume without paying tribute to all those who contributed to this momentous project and to the cooperation that grew between British and French engineers as they confronted the same problems together. Everybody who ever worked on the Tunnel should share in a collective pride in the completion of the most important civil engineering project this century. Engineers love designing and building things, they see their various constructions as their reward and memorial; they are generally not keen on writing about what they have done. Those who offered to assist in the preparation of this brief record of their achievement are particularly to be thanked. We all trust that this book has conveyed a sense of the scale, complexity and sheer excitement involved in engineering the Channel Tunnel.

Contributors

Colin Kirkland has a lifetime's experience in the tunnelling field. He has represented the UK in the International Tunnelling Association since 1986, and held the office of President 1989–92. He was Chairman of the British Tunnelling Society 1985–87. In July 1991 he was elected a Fellow of the Royal Academy of Engineering. A member of Halcrow's core tunnelling staff since 1952, in 1986 he began a six-year secondment to Eurotunnel, where, as Technical Director, he was involved in financing, management preparation, design and construction of the Channel Tunnel. He had a special responsibility for external relations worldwide as well as environmental management of the UK works. In recognition of his work on the Channel Tunnel he was awarded the OBE in 1992. Colin Kirkland's major contribution to tunnel engineering and his expert understanding of the roles of consultant, contractor and employer have evolved over three decades of management of major projects. These include the supervision of construction of underground railway systems in London and Glasgow, and many miles of communications tunnels in London, Birmingham, Manchester and Cardiff. He is now a member of the Management Board of Sir William Halcrow and Partners Ltd responsible for Company Development.

Michael Baxter was Project Engineer, Channel Tunnel Trackwork 1990–1992 and responsible for: track design of crossovers (land and undersea); resilience of turnouts at crossovers; coordination of track design with all interfaces, e.g drainage, air lock doors, ground beams, walkways, mechanical equipment, electrical equipment, ducts and cables; draft reports on results of laboratory tests on the Sonneville track system (for main rail tunnels); review and comments on tracklaying contractor's method statements and programme; design coordination with Terminal trackwork for transition structures at tunnel interface and special-purpose sidings. He commenced his career with British Railways and subsequently worked in Kenya (EAR); South Africa (Coal Line Project to Richards Bay, 1972–77); and Singapore MRT as Senior Track and Alignment Engineer during construction of the new metro system, 1984–1990.

Alain Henri Bertrand is Eurotunnel's Deputy Director-General, Operations, heading the division with responsibility for running the Channel Tunnel transportation system, and for the human resources, new works, purchasing and other departments. He joined Eurotunnel in 1987, after working with SNCF for more than 20 years, proceeding from the marketing department to Chef de Gare at Paris-Austerlitz, then Commercial Director, Passengers, in the Rhône-Alpes region to Head of the Transport Division and Deputy Director in the Provence-Alpes-Côte d'Azur region.

Michael A C Cowan, FIMechE, MCIBSE is a Technical Director of W S Atkins Consultants Ltd. He worked on various mechanical and electrical aspects of the Channel Tunnel, initially in the Maître d'Oeuvre organisation and latterly on secondment to the Transport System Division of Eurotunnel, up until completion in 1994.

John Davies is a Senior Consultant in the Transport Division of W S Atkins and has been involved in the Channel Tunnel project since September 1986. He was initially part of the Maître d'Oeuvre team, but was transferred on secondment to the Operations Department of Eurotunnel in late 1987. He has since worked as Terminals Liaison Consultant under the directors of the UK and French terminals. After involvement in the overall design of the terminal layout, he was Operations representative on the Eurotunnel project team dealing with the Terminal Traffic Management contract. He has had the advantage of following this contract from initial stages right the way through to implementation and commercial operation culminating in the recent introduction into service of the upgraded SEMA toll system.

Peter Davies was Consulting Engineer to the Trackwork Division of TML, 1988–95, his responsibilities covering all aspects with particular emphasis on the UK terminal. Included were: track foundations; drainage structure gauge; gauging maintenance access and procedures; trackwork alignment; detailed design of trackwork (directly supplied materials and systems including rails, turnouts, points heating and lubrication); subcontracting supervision. Previous to the Channel Tunnel, he worked for many years in Africa, initially with Rhodesia Railways (1948–1957), then with a civil engineering consultancy (1957–87) of which he became Senior Partner and Group Chairman. During this period he was responsible for some 1500km of railway route location, design and construction, including in many cases complete infrastructure of townships, workshops, communications, signalling, maintenance facilities etc., plus the investigation and design of seven tunnels.

Roger Ford MCIT, AIRSE trained as an engineer with English Electric and on qualifying joined the company's Railway Division. Subsequently he pursued a career in industrial publicity management before, in 1976, becoming a full-time technical writer specialising in railways. He is industry and Technology Editor of the magazine *Modern Railways* and also edits the newsletter *Rail Privatisation News*. He is the co-author of HST at Work, the standard work on the development and operation of British Rail's InterCity 125 fleet, and is a regular contributor to technical magazines in the UK and abroad.

John Finn was Power Supply Project Manager with TML on the Channel Tunnel project. He trained with the Central Electricity Generating Board, initially in power stations and subsequently in transmission. With the supply industry, he worked in operation maintenance, protection and system planning before joining private industry, where he worked on several EHV overseas projects for British and Japanese contractors. An IEE Fellow, he is now Chief Engineer for Reyrolle Projects.

Tim Green BEng CEng FICE joined the Channel Tunnel project in late 1989 as an experienced railwayman. For Translink Joint Venture, the project contractor, he was responsible for the narrow-gauge construction railway, the standard-gauge permanent railway and other logistical services used to support construction. He also controlled the pithead site at Shakespeare Cliff and was responsible for the management and maintenance of construction locomotives, rolling stock field plant and a fleet of road vehicles. He led the joint management and trades union co-operation on the management of site safety.

Roger Hacquart, Ingénieur de l'Ecole Supérieure d'Electricité, spent 25 years in the steel industry before moving to railways. He joined Eurotunnel, where he was heavily involved in the mechanical and electrical design of the fixed link, before becoming Technical Director. He is now in charge of relations with the national railways, the most important customer of Eurotunnel.

David A Henson PhD began development of a method of analysing the pressure transients in single tunnels as part of his doctoral thesis in the late 1960s. This work was continued for British Rail to develop the theory for the three-tunnel Channel Tunnel configuration, as then proposed. He joined Mott Hay and Anderson (later Mott MacDonald) in 1973, where he is now a Divisional Director in the Tunnels and Highway Services Division. Here he continued to develop the conceptual design of the Channel Tunnel, expanding its scope to include the normal and emergency ventilation and the tunnel cooling. Over this period he has also been involved with the ventilation, draught relief and cooling of more than 30 main-line railway and rapid transit projects worldwide.

Karel de Jaeger-Ponnet is an international expert in telecommunications with the Transportation Department of Tractebel Development, Brussels. He was Project Manager, Telecommunications, within Eurotunnel's Project Implementation Division, responsible for: follow-up of design and implementation of the data transmission system (including digital transmission on optical fibres, telephone, public address and clock systems) and of the radio systems; obtaining UK and French operating licences for this system; liaison with BR and SNCF, and the UK and French authorities, on normal and emergency communications; technical assistance to the Commercial Department regarding the international telecommunications link through the Tunnel, terminal payphones and other data/telephone services.

Guy A Lance BSc CEng MICE, currently Head of Tunnelling at WS Atkins, is a practising Engineer with more than 25 years' experience in the design of underground works. He was first involved in the Channel Tunnel project in 1974, and then again in 1985, when he was part of a team carrying out a Technical Audit of Channel Tunnel Group's submission to the two governments. Following award of the Concession he undertook responsibility within the Maitre d'Oeuvre for UK tunnel design before moving across to Eurotunnel as Project Design Manager, Tunnels UK. In this role he led the tunnelling through to completion in 1991.

Yves Machefert-Tassin is a specialist in electric motive power and has been involved in designing many locomotives and power units. He is the author of several books on the evolution of electric and diesel motive power, and is a regular contributor to *La Revue Générale des Chemins de Fer* and other French railway journals.

Bob Marshall is a Chartered Civil Engineer with over 18 years' experience in the design and construction of tunnels in both the UK and overseas. He is currently employed as a Principal Engineer in the Tunnelling Department of Howard Humphreys/Brown and Root Civil. During his involvement with the Channel Tunnel, he was seconded to Eurotunnel from Sir William Halcrow and Partners Ltd. From May 1989, he was Assistant Section Manager on the UK Marine Rail Tunnel drives and took over, from mid-1990, as Section Manager for the Marine Service Tunnel, UK Crossover Cavern, and all marine hand tunnel work. This involvement with the project continued on a full-time basis until mid-1994.

Richard Morris, following management training in 1970 with British Rail, worked in several posts mainly concerned with Operations, including Area Manager, Paddington, and Area Manager, Tinsley (Sheffield). He was appointed Operations Manager for the whole of British Rail in 1990, with additional responsibility for security matters. In March 1992 he joined Eurotunnel with particular responsibility for compilation of the Rules by which the railway and terminal system would operate. He was appointed Safety Director in January 1993, reporting to the Chief Executive, and responsible for the implementation of Eurotunnel's Health and Safety policy and safety training, the development of the company's safety management system, and also monitoring and providing all necessary support and advice to safety managers throughout the company. From November 1994, the Directorate also encompassed Quality and Occupational Health.

Eric Radcliffe, after being Project Surveyor for the uncompleted Channel Tunnel construction 1973–75, was Chief Surveyor for construction of the UK Tunnels 1987–91, with control of all drive alignments for the achievement of successful junctioning. Trained as a surveyor at college, in the Royal Artillery and at Ordnance Survey, worked as a civil engineering surveyor for local authorities in the UK, and on construction projects in Africa and the Middle East. A Fellow of the Institution of Civil Engineering Surveyors, he was awarded the MBE in the New Year Honours List, December 1991, for services to civil engineering.

Paul Michael Robins CEng MIEE MIRSE, after an Electronic Engineering degree at Southampton University and seven years in the Signal Engineering Department at London Underground, joined W S Atkins and almost immediately was seconded to the Channel Tunnel project, initially to the Maître d'Oeuvre and then to the Project Implementation Division (PID), where he was Signalling Project Manager. Responsible for agreeing technical specifications and monitoring progress of the signalling system with TML and its subcontractors, he also negotiated with BR and SNCF and with the Safety Authority to obtain the necessary approvals. Since 1992 he has been employed by Eurotunnel as Control and Communications Manager in the New Works Division, based in the Siège in Coquelles, France.

Peter William Brett Semmens MA CChem FRSC MBCS MCIT FRSA has a degree in Chemistry from Oxford University, and worked in the chemical industry for 25 years before, in 1974, becoming the first Assistant Keeper of the new National Railway

Museum at York. He has been author or part-author of 30 books, and for 14 years has written the monthly 'Railway Practice & Performance' article in *The Railway Magazine*. In 1990 he was appointed Chief Correspondent of *The Railway Magazine*. His extensive writings about the Channel Tunnel include two books on the subject.

Martin Stearman Dip Arch RIBA BTp MRTPI was seconded from WS Atkins into the Eurotunnel project from 1986 to 1994. Chief Architect/Planner in the Maître d'Oeuvre organisation 1986–88, he later became Design Manager UK Terminals in the Project Implementation Division. He was Member and Secretary of the Eurotunnel Corporate Identity Working Group 1988–94 and Environmental Coordinator UK Terminals 1991–94. His career prior to the Channel Tunnel project included redevelopment of London's Piccadilly Circus and Canary Wharf, with work abroad in Hong Kong, Venezuela and Nigeria. He is currently Managing Director of Atkins Lister Drew Limited, the specialist Architectural Division of the WS Atkins Group of Companies.

Paul Varley was Geotechnical Design Manager for TML on the Channel Tunnel. He joined TML in 1986 and remained on the project until 1992 throughout the site investigation and construction of the works. He has supervised the design and construction of a wide range of underground works, including the 30m-span power cavern at the Pergau Hydroelectric project in Malaysia. Dr Varley is currently the Head of Knight Piésold's Tunnelling and Rock Mechanics Group, responsible for the design, specification and construction supervision of all underground works associated with the firm's power, mining, water transfer and transportation projects. During his career Dr Varley has undertaken site investigations, planning and design work, structural surveys for tunnels, instrument installation and monitoring, and the formulation of contractors' claims.

David C Wallis MICE is an Engineering Manager in the Tunnels Department of Sir William Halcrow and Partners Ltd with 31 years' experience in tunnelling projects. He has been responsible for the conceptual design of road and metro schemes in the UK and abroad and has supervised major tunnelling projects in soft rock, clay, and waterbearing gravels under compressed air. Between 1986 and 1991 he was responsible for the firm's involvement in all tunnelling and related aspects for the UK side of the Channel Tunnel works, firstly in the Maître d'Oeuvre team and then as Tunnel Project Manager seconded to Eurotunnel with a staff of over 50 Halcrow personnel, checking design, monitoring construction, procurement and progress, and auditing cost. He was subsequently involved in the conceptual design of the UK underground nuclear waste repository at Sellafield in Cumbria, and advised on the design of the Guangzhou metro in China, the Gibraltar Strait Crossing and a number of smaller projects. In late 1994 he prepared the tunnelling proposal for the London–Folkestone high-speed rail link.

Colin David Warren BSc (Geology) MSc (Soils) DIC MIMM CGeol CEng is a Senior Geotechnical Engineer with Sir William Halcrow and Partners Ltd and has over 23 years' experience related to the geotechnics of major civil engineering projects both in the UK and overseas. Between 1987 and 1992 he was seconded to Eurotunnel as Chief Geologist, with special responsibility for monitoring all geotechnical aspects associated with UK construction of the Channel Tunnel. Other typical projects include the Dubai Dry Dock, Great Belt Scheme (Denmark), Gibraltar Straits Crossing, the Folkestone Warren landslides and more recently the proposed high-speed link between London and Folkestone. He still lives in Folkestone and is married with three children, his youngest being named Thomas Beaumont in recognition of Colonel Beaumont, whose 1881 tunnel at Shakespeare Cliff was finally sealed by grout on the day of his birth in October 1988.

Eric Whitaker QFSM FIFireE FIRM, a UK Fire Service Officer for over 30 years, has served on national and international committees concerned with operational firefighting and research issues. He was President of the Institution of Fire Engineers and a member of its Council for ten years. In 1984 he was appointed as Technical Advisor to the Institution and in 1986 transferred to the Channel Tunnel Group as its principal fire safety adviser. Since then he has advised on design, test and research matters with fire safety implications for the Tunnel.

Selected abbreviations

ATP	automatic train protection	NATM	New Austrian Tunnelling Method
BR	British Rail	OBC	operations board controller
CML	Continental main line	RCC	railway control centre
CR	Concession radio	RMC	rail movement controllers
CTG	Channel Tunnel Grid	RT	rail tunnel
CTG/FM	Channel Tunnel Group/France Manche	RTM	railway traffic management system
CTTG	Channel Tunnel Trackwork Group	SA	Safety Authority
EdF	Electricité de France	SIR	shuttle internal radio
ET	Eurotunnel	SNCB	Société Nationale de Chemins de Fer Belges
EMS	engineering management system	SNCF	Société Nationale de Chemins de Fer
HGV	heavy goods vehicle	ST	service tunnel
IGC	Intergovernmental Commission	STTS	service tunnel transport system
LADOG	light service tunnel vehicle	SUD	Shakespeare Underground Development
LPG	liquid petroleum gas	TBM	tunnel boring machine
LRTN	land rail tunnel north	TCC	terminal control centre
LRTS	land rail tunnel south	TCSC	traffic control and supervision computer
LST	land service tunnel	TCT	toll collector terminal
MC	marshalling console	TGV	*train à grande vitesse*
Md'O	Maître d'Oeuvre	TML	Transmanche-Link
MIS	management information system	TSC	toll station computer
MRTN	marine rail tunnel north	TTM	terminal traffic management
MRTS	marine rail tunnel south	TTR	track-to-train radio
MST	marine service tunnel		
MTC	main toll computer		

Index

1973–75 project 1, 23–24, 40, 51, 64, 79–80, 85, 88–89, 91, 105

A20 road 86, 261, 269
Abbots Cliff tunnel 79
access 15, 65, 67, 79–90
acoustic screen 266
adits 40–41, 54–55, 67, 81–84, 118, 154
adjusted schedule 154
ADP 262–263
ADSA 262, 265
aerodynamic drag 211, 216–218, 222, 225–226, 244
aerodynamics 16, 211, 216–226
air bags 197–198
air brake 198
air conditioning 179, 183, 199–200, 202, 215, 226
air handling units 213, 215, 219
air pressure 193, 211, 214, 216–219, 221–222, 224–226, 244
air velocity 218–219, 221–222, 226
alignment 55
alignment transfer 54
alkali silica reaction 78
aluminothermic welding 171
ambulance 248
amenity coach. See club cars
amenity facilities 257
analytical models – linings 70
ANF Industrie 191
antiwheelslide 208
archimedean screw conveyor 70
arranging banks 3
ASEA Brown Boveri (ABB) 175, 177–179, 184
Ashford 87, 155, 268, 297, 300
astronomic azimuths 52
asynchronous drive 189
asynchronous motor 177
atmospheric refraction 51
autocouplers 179
automatic ticket readers 275

automatic train protection (ATP) 149–150, 302, 305,
auxiliary power system 140–141
'avant projet' 11–12
axleboxes 178, 183, 197, 207
Aycliffe housing estate 85
azobe timbers 159, 165–166

backup gantries 92
Balfour Beatty 3, 161, 243
ballasted tracks 157–159, 161–162
Banque Indosuez 3
Banque Nationale de Paris 3
baseline 54, 55
basement rocks 25
battery-backed fittings 143
BDP 262–263
bearing pads 93
Beaumont tunnel 22, 40, 64, 79, 84
bellows 193
benchmark bolts 57
bentonite cement grout 31
Beussingue portal 90, 161
Bigginswood 267
block sections 150–151
Bo-Bo wheel configuration 177, 181
bodyshells 178, 183
bogies 177–178, 180, 183–185, 187, 189–190, 197–198, 205–206, 208
bolted linings 92, 105
boltless linings 92–93
bomb disposal 160
Bombardier Eurorail 191
Bombardier Prorail 183
'boot factory' 167, 170
boreholes 22–26, 58, 91–92, 96, 98
Borie SAE 164
Bouyges 3
British Rail (BR) 130, 153, 166, 175, 181, 183–184, 186–187, 190, 209, 261, 295–297
BR train radio 130
brakes 198–199, 208

breakthroughs 37–39, 58, 60, 90–91, 107–108, 114–115
Breda-Fiat 205
brickpit structure 266–267
bridging plates 196, 206
British Rail Research 218
Brush Traction 175, 177, 183, 187
'build bars' 101
butterfly dampers 213
Buxton HGV fire tests 237–238

cab signalling system 149
Calais Terminal 13, 131, 135–137, 140, 146–148, 157–161, 245, 255, 257, 263, 272, 280, 292
camper vans 234–235, 278
cant 173
Canterbury Archaeological Trust 269
capacitive bank 138
car parking 86, 283
caravans 196, 234–235, 278
carbon monoxide detectors 238
carbonate mudstone 41–42
Cardington airship hangars 234
carrier frequency 151
Casagrande 165
cast-iron linings 66, 68, 72–73, 76, 79, 88, 91, 99, 109, 220
Castle Hill 25, 28, 30, 40, 43, 68, 80, 94, 132, 161, 261
catenary 141, 147–148, 176, 190, 211, 217, 220, 226, 238, 296, 311
chalk marl 24, 26–29, 34, 38, 41–42, 44, 46, 48–49, 79, 81, 90–92, 96, 108, 112, 120, 261
Channel Expressway 5
Channel Tunnel Bill 6, 81, 83, 267–268
Channel Tunnel Grid 51–52
Channel Tunnel Group/France Manche. 3, 80, 297 *See also* Eurotunnel
Channel Tunnel Height Datum 53
Channel Tunnel Trackwork Group (CTTG) 164, 166

Charles de Gaulle Airport 256
check rails 160
chef de train. *See* train captain
Cheriton. *See* Folkestone Terminal
Chief Inspecting Officer of Railways 231
Chief Inspector of Fire Services 231
Chinnor chalk-cutting trials 91
Class 120 locomotives 177
Class 30 locomotives 177
Class 450/460 locomotives 181
Class 60 locomotives 183
Class 92 locomotives 175, 181–189, 297–299
Class BB22200 locomotives 297
closed polygons 56
closed-circuit television system 175, 201
club cars 203, 205–206, 208–209, 236–238
Co-Co wheel configuration 177, 298
coaches 191, 196, 234, 295
collision forces 179
Comatelec 243
commissioning tests 221
communications 16, 129, 142, 146, 154, 175, 179, 200, 248, 305
Conbex 802 retarder 103
Conbex 803 accelerator 103
Concession 2, 6, 63, 155, 203, 241, 301
Concession radio (CR) 129–132
concrete 78, 87
'concrete bullets' 119, 121
construction railway 117–124, 166
consultation 268–269
contact shoes 185
Continental main line 155, 160, 257, 266
contract 9, 86, 272
  lump sum 9
  target cost 8
contractors 3
control centres 129–132, 147, 152–155, 189, 215, 238, 245, 294, 305

control system 16, 153–155
cooling 16, 120, 140–141, 154, 211, 217–218, 220, 226–229, 294
Coquelles. *See* Calais Terminal
corporate identity 262–265
corrosion 145, 164
cost 314
cost control 9–10
Costain 3, 165
craie bleue 28
credit cards 272, 276
Crédit Lyonnais 3
Cretaceous 21, 26
cross-channel cable 266
Cross-Channel Contractors 91
cross-passages 14, 30, 35, 49, 65–66, 72–73, 121, 141, 152, 172, 209, 213, 215–216, 219–221, 224, 226, 241, 247, 291
crossover caverns 14, 30, 41–43, 65, 69–70, 123, 244, 291, 303
crossover doors 219, 224, 226
crossovers 14, 25, 40, 113, 121, 123–124, 132, 161, 165, 168, 219, 221, 303
culbuteur 90
cutting heads 57–58, 66, 70–71, 93, 97–98, 101–102, 104, 108

dampers – piston relief ducts 141, 215, 219, 221, 224, 236, 303, 306
Danton Pinch 266
DB/Rheda 162
De Dietrich 193
de Gamond, Thomé 21
Déclaration d'Utilité Publique 6
decoding functions 152
deconfinement 75
derailment 183, 304
descenderie 24, 89
Deutsche Bahn (German Federal Railways) 177, 295
diameter of tunnels 8
diamond crossing 165
diaphragm walling 89
diesel engines – STTS 213

diesel locomotives 119, 311
diesel multiple units 125
'Diplodocus' 169
disabled passengers 196, 202, 234
disc cutters 98
Disney 263
Disneymobile 123
Dollands Moor 80, 155, 161, 268, 300
double-deck carrier wagons 191–194, 197–198, 277
double-deck loading wagons 196
downhole geophysics 25–26
downholes 49
draftgear 179
dragging equipment detectors 304
drawhook 179
dual-voltage operation 185
Dumez 3
duty free facilities 257
'dynamic' encoded tags 275

earthing system 134
earthquakes 12, 25, 303
Echingen electricity station 136
Edilon Corkelast elastomer 165
effective stress case 75
electric locomotives 118
electrical rooms 140, 142, 145
electrical substations 144–145, 238
Electricité de France (EdF) 135
electromagnetic interference 180
embedded rails 159
emergency siding 160
emergency situations 173, 176, 182, 194, 211, 215–216, 232, 236, 238, 241, 245, 266, 291, 303, 311
EMS functions 154
engineering management system (EMS) 153, 245, 294
Environmental Forum 268
Environmental Impact Assessment (EIA) 267–268
EPDC 242
equipment rooms 151, 221, 238–239, 247, 306
error checking 151

escarpment 39, 43
ETR 500 208-209
Eureka project 237
Eurobridge 5
European Passenger Services 182–183, 295, 298
EuroRoute 5
Euroshuttle Locomotive Consortium 177
Euroshuttle Wagons Consortium (ESCW) 191
Eurostar 14, 155, 157, 175-176, 189–190, 197, 292, 295, 298-299
Eurotunnel 3, 63, 86, 108-109, 114, 149, 155, 175, 177, 187, 191, 203, 225, 231, 234, 236, 241, 255, 267, 272, 284, 291-292, 295, 297, 301, 303, 315
Eurotunnel Exhibition Centres 115, 268
'Eurotunnelworld' 263
evacuation 132, 143, 146, 154, 172, 174, 194, 202, 209, 211, 215-216, 232, 236, 238, 241-242, 245, 247, 306-308, 310-311
Fabeg electrical heads 179
fail-safe display 150
false alarms 202
fan building 214
fan layout – track 160
fans 214-216
Farthingloe 86
fault level 136
fault level (electrical) 135
fault zones 91
faulting 21, 34, 37
Fiat Ferroviaria 205
fibre-optic cables 129
fingers-trailing 107-109, 114-115
fire
　alarm 201-202, 232
　dampers 200, 202
　detection 189, 201-202, 209, 232-239, 307
　extinguishers 181, 186, 189, 202, 232-233, 235, 239, 307
　load 234
　prevention 231-232
　procedures 236-237
　protection 96
　resistance 232
　safety 16, 143, 145, 154, 181, 185, 187-189, 192-193, 201-202, 209, 215-216, 241, 243, 245, 266, 291, 294, 298, 306-309
　sensors 202
fire-resistant (micc) cable 243
firefighting 140-141, 146, 160, 248-249
first contact MST 107
flatbed wagons 101, 103-104, 119, 121-122, 124, 170
flexible coupling 177
fluid dynamics 215, 234
folding 34, 37-38
Folkestone Beds 44
Folkestone Terminal 13, 68, 124, 131, 134-135, 140, 146-148, 157-161, 167, 176, 245, 255, 271-287, 300
Fond Pignon 48, 90
Fosroc grouting system 103
Fosse Dangaerd 25, 28, 30, 37
Framafer RND 92 173
France
　access 67, 88-90
　concrete joint tests 76
　control centre 153, 294
　crossover 69, 165
　fan 214
　geology 29, 48, 90
　grid connections 135
　lining factory 90, 165
　linings 70-71, 75
　main substations (electricity) 136-137
　marine and land tunnels 90
　marshalling tunnels 89
　openings from service tunnel 73
　power system 133-134
　pumping station 140
　rail tunnels 69, 72-73, 169
　service tunnel 88
　signalling 155

toll system 273
tracklaying 166, 169
tunnel boring machines 75, 94, 114
ventilation building 214
free exit 257
freezing 229
freight shuttles 157, 203–209, 218, 236–238, 277, 295–296, 308
freight trains 157
Fréthun 155, 182, 186, 295, 299
friction factor 220
frontier controls 257, 276, 277
fumes 199–201, 209
funding 7–8

gantries 101
gantry cranes 171
Gare du Nord, Paris 295
gases 202
gate turn-off (GTO) thyristors 180, 184
gauge 173
gauge bars 171, 173
gauge cutters 98
gault clay 24, 26–28, 31, 38–39, 41–44, 46, 49, 81, 96, 159, 161
gearbox 177
GEC Alsthom 176, 189, 191
geological succession 27
geology 14, 21–49, 63–64, 90, 97
geophones 26
geostatistical methods 26
German Federal Railways (Deutsche Bahn) 177, 295
glauconitic marl 26, 28, 31–32, 37–39, 41–42, 44, 46, 107
Glensanda 87
Global Positioning System 53
GMT-Bombardier 191
Goodwin Sands 265
gradient 151, 184
'greasybacks' 39
grid connections 134–136
grippers 93, 96, 101–102, 104, 109, 173
gross tonnage 157

'ground model' 74
ground treatment 31, 90–91, 97–98, 114–115
grout curtain 89
gyrotheodolite 55–57

Halon 1301 181, 186, 189, 233, 239, 307
Hammerfest HGV fire tests 237–238
hazardous goods 235, 309–310
headways 149, 217, 236, 292
Health and Safety Executive testing station 237
heat exchange 228
heat gain 227
heat removal 227–228
heavy goods vehicles (HGVs) 236–238, 277, 295
height limit detectors 275
Heitkamp 164
Holywell 43, 45, 58, 68, 80–81, 86, 115, 132, 165
Hong Kong rapid transit system 226
hoods 39, 96, 106, 108
hood canopies 196
'hot standby' 294
hotbox 304
House of Commons 269
House of Lords 81, 269
Howden TBMs. see James Howden and Co.
humidity 228
'hunting' 198, 208
hydraulic dampers 197, 207–208
hydrofracture 36, 112
Hythe 265

IEE Regulations 97
immersed tube tunnel 24
impedance 135–136, 138–139
infrared flame detectors 237
Institut Géographique National 51–53
insulation 192
interactive behaviour 75

InterCity 209
Intergovernmental Commission (IGC) 10–11, 231, 235, 265–266, 277, 285, 301
interlocking 153
International Union of Railways (UIC) 297
inundation 94, 96, 101–102
ion concept 234
ionisation detectors 237
IP 65 244
IPA 162
Isle of Grain 78, 86–87, 118
Italian State Railways 208

James Howden and Co. 100, 104, 112, 115
Jasmine computer model 234
Joint Consents Team 268
Joint Consultative Agreement 268
Jurassic 26

Kent coalfield 26, 79, 265
Kent Fire Brigade 232, 237
key ram 103
key word descriptors 262
Kidde Fire Protection 189
knife seals 96

La Brugeoise et Nivelles 191
LADOGs 248—251
lagoons 84–85, 265
landscaping 262, 267
'landside' 257
landslip stabilisation 45–46
laser beams 57–58, 98
lateral force 218
Le Shuttle 14
learning curve 108
Les Attaques electricity station 136
Les Mandarins electricity substation 135–136
life protection 236
light service tunnel vehicle. *See* LADOG

light-emitting diodes 241
lighting 141, 143–144, 146, 205, 226, 241–245
linings 14–15, 29, 40, 63–78, 92–93, 98–100, 106, 109, 114
  design 63–78, 80
  durability 78
  factory 78, 86–87
  rings 36, 71–72, 87, 93, 95, 121
  segments 66, 70, 103, 115, 117–118, 121, 220
  testing 87
liquefied petroleum gas (LPG) 232, 235, 278
listed buildings 269
live animals 309
loading plates 196–197
loading/unloading wagons (shuttle) 191–192, 195–203, 205–209, 238, 278
Loco 2000 184
Logica 271
logistics, construction 15, 117–126
London Underground 284
London-Channel Tunnel Rail Link 176, 295
Longport House 269
loop tunnel 266
Lötschberg tunnel 302
low-voltage power system 141–144
lower greensand 26, 44
lubricators 160
luminaires 242–245
Lydden Fault 38

M20 motorway 80, 86, 155, 256, 261, 266–267, 269
Maggiemobile 123
main electricity substations 136–139
main toll computer (MTC) 274, 276
maintenance 211, 219–220, 238, 241–242, 244, 248, 250, 292–294
Maître d'Oeuvre 4, 10, 108, 233–234, 236, 241

major incident coordination centre 294
management information system
 (MIS) 274
manriders 119, 122–123
marche à vue 305
marker lights 241–242
marshalling areas 55, 67–68, 83, 101
marshalling consoles (MC) 280
master switch controller (MSC) 131
maximum train load 179
mechanical cooling system 226–228
microfossil analysis 24, 29, 31
Midland Bank 3
Mill House 269
mimic board 124
mimic diagram 154
Mines Research Establishment 233
minestone 265
minibuses 191
monocoque body 178–179, 183
monoprocesseur codé 151–153
Montcocol 164
motorcycles 196, 295
Motorola Storno 130
Mott MacDonald 220
Mount Baker Ridge freeway tunnel 69
mountain bicycles 126
muck skips 119, 121–122
Multifret wagons 296

Nabla rail fastenings 159
national railways 149, 155, 296–300
National Westminster Bank 3
negative phase sequence (NPS) 135, 138
neoprene seals 71, 73
Netherlands Railways 295
network code 151
New Austrialian Tunnelling Method
 (NATM) 30, 37, 39–49, 67–
 70, 73, 115
New Zealand Railways 177
Newington 257, 266
Nivellement Trans-Manche 1988 54
nonballasted trackform 161
nonreturn dampers 213, 219

nonsegregation concept 231–232
Nordrach 249
normal ventilation system 212, 215, 219
nuclear flasks 309

Observational Approach 40
occupied block section 150
'one-pass' support 66
'open days' 268
operation of Tunnel 291–300
operations board controller (OBC) 124
optical detectors 237, 238
Ordnance Datum Newlyn 53
Ordnance Survey 51
organ pipe effect 218
outages 140
overbreak 33–39, 109–110, 115
overbridges 260
overhead traction system 134
overlap block section 150

PACT 162
Palaeozoic 26
pantographs 138, 180, 185
Paris Métro 284
pass doors 194, 202
passenger terminal building 262, 276–
 277
passenger-vehicle shuttles 157, 175,
 191–202, 231, 295, 307 *See also*
 shuttle locomotives, shuttle wagons
Peck 40
Peene 257, 266
permeability 30, 37, 41, 67, 70, 75, 78,
 88, 90, 109, 112
permissive blocks 150
petroleum vapour 232
Phoenix 165
picks 98, 101–102, 104, 108
pictograms 286
piston discharger 70
piston effect 211, 226
piston relief ducts 14, 30, 35, 49, 65–
 66, 72, 121, 200, 211, 213, 215–220,
 222, 225–227, 244, 309

pithead   120
planning process   267–268
platform tracks   159, 160
platforms   260, 279–280, 286
pods   238, 248
point heaters   166
'point M'   166
porewater chemistry   33, 36
power system   15, 133–148, 154, 175–176, 185, 190, 211, 217–218, 220, 225–226, 244, 304
Pozament GP3   103
PRCI   153
pressure pulses   179, 219
pressure relief ducts. See piston relief ducts
primary stations   54
primary subsystems (power)   134
primary suspension   178
probing   31–32, 41, 49, 94, 98
program cycle   153
programmable logic control (PLC)   111
Project Implementation Division   11
project standards   77
propel rams   93, 96, 99, 102, 104
pumping stations   65, 68, 140, 144–146, 154
purging – fumes   199–200

Qualter Hall   178
Quenocs anticlinal fold   27
quill tube   177

rack and pinion railway   118
radio systems   16, 129–132
rail life   161
rail movement controllers (RMCs)   124–125
'rail string'   170–171
rail traffic management system (RTM)   152–153, 280, 294
Railfreight Distribution   182–183
Railtrack   155
raise bore technique   30

rakes   191, 203, 205
rapid transit systems   226
reference buildings (terminals)   256
Reflex Maxibor downhole instrument   59
refrigeration plants   227–228
regenerative braking   184
reinforced earth   266
RER Line A (Paris)   152
Réseau du Tunnel sous la Manche 1987 grid   53
reservoir lanes   281
resilient pads   165
resistance to fatigue   225
retractable shoegear   190
rheostatic braking   184
Robbins Markham Joint Venture   100
Robbins/Markham TBMs   102
rock mass quality   41
rockbolts   40
'rolling motorway'   278
Round Hill   86
route-setting   153–154
Royal Fine Arts Commission   268
rubber boots   164–165, 170, 304
rubber-tyred vehicles   117, 125, 247–248

Saar Gummiwerke   165
SACEM signalling system   152
safety   96, 143, 185, 231–239, 241–245, 297, 301–311
Safety Authority   11–12, 17, 64, 76, 184–185, 231–238, 241, 251, 301
Safety Case   231, 302
salinity   42
Sangatte   24, 48, 89–90, 131, 140, 146, 165, 214
satellite location   52
Scharfenberg coupler   179
Schweizerische Bundesbahn (Swiss Federal Railways) 177, 181
Schwing pump   103, 107
screw conveyor   90
sea wall   83–85, 120, 265
sealed cabs   179

secondary suspension 178, 183
security inspection 258
segment erectors 101, 103–105, 111–112, 115
Seikan tunnel 63, 297
seismic profiling 25
seismic reflection survey 24
seismic traverses 25
Select Committee 6, 256, 269
Sellindge electricity substation 134–135
SEMA 271, 273
semi-open wagons 203
Shakespeare Cliff 24–25, 30, 40, 52, 54, 67, 79–81, 83, 85–87, 94, 97, 114, 117, 120, 140, 146, 167, 214, 265, 268
Shakespeare underground development (SUD) 30, 40–41, 68
shinkansen 302
shotcrete 40, 43, 114
shunting controls 206
shuttle doors 225
shuttle internal radio (SIR) 129, 132
shuttle locomotives 175–181, 191, 203
shuttle loop 160–161, 176
shuttle wagons 191–209, 218, 277 *see also* double-deck wagons, single-deck wagons
shuttles 149, 157, 231, 295, 299 *see also* freight shuttles, passenger vehicle shuttles
Siemens Plessey 271
signalling system 16, 149–153, 155, 157, 189, 201, 241, 299, 305
signs 262, 264–265, 271, 278, 283–287
Silacsol T grout 35
Simplon tunnel 177, 225
Singapore rapid transit system 226
single traverse 56
single-deck carrier wagons 191–192, 194–195, 197–198, 277
single-deck loading wagon 196–197
single-heading 184
single-line working 219–220, 222, 244
single-track working 293

Singleton 269
Sites of Special Scientific Interest 43, 68, 81
sleepers 157
slip forming 68
slurried chalk spoil 90
smoke 145, 200–202, 209, 215–216, 232, 234, 236–239, 266, 291, 302–303, 306
SNCB 175, 190, 295
SNCF 150, 153, 155, 159, 166, 175, 181, 183, 186, 190, 295, 296, 298
SNCF train radio 130
Société Auxiliaire d'Entreprises 3
Société Générale d'Entreprises 3
sodium silicate chemical gel 31
solid-state interlocking system (SSI) 153
Sonneville track system 162–165, 170
speed code 151
Spie Batignolles 3, 161, 243
spiles 46
stabilisers 194–195, 205–206
standard-gauge railway 117, 124–126
standby generators 146
static and dynamic elasticity 163
steer shoes 104
Steinmetz principle 138
stereonets 30
stock rails 166
Stone Farm 269
stored value 272
structural measuring stations 78
structure gauge 178
STTS 17, 125, 129, 241–242, 247–248, 309
STTS vehicles 241–249
substations 140–142, 147
Sugar Loaf Hill 30, 68, 108
supplementary ventilation system 212, 214–216, 237
survey closure 58–59
survey stations 55
surveys 14, 23–26, 51–60
Swiss Federal Railways (Schweizerische Bundesbahn) 177, 181

switch rails   166
switchblades   166

tactical radios   129
tactical schedule   154
tail seal   71
tailskin   29, 70–71, 92, 96, 98, 105–107
target cost contract   9
target speed   151
tariff bands   272
tariff categories   272
Tarmac   3, 164
Taylor Woodrow   3
temperature   226, 228–229
Terminal 4, Heathrow Airport   256
terminal traffic management (TTM)   271–287
terminals   13, 17, 132, 166, 255–269, 271–287. See also Calais Terminal, Folkestone Terminal
testing – linings   87
TGV Nord Europe line   150, 161, 189–190, 295, 305
TGVs   175, 189–190, 197
third-rail power supply   176, 181, 184–185, 190, 296
throws   27
thrusting rams. See propel rams
thyristor-controlled reactor (TCR)   138
timetable   154, 299
TML   3, 6, 9, 52, 91–92, 108–109, 114, 163–166, 233–234, 241–242, 267–268, 271
toilets   192, 194, 196–197, 209
toll booth   274
toll collector terminal (TCT)   273–275
toll lane   274
toll operator   274
toll plaza   274
toll station computer (TSC)   273, 275–276
tolls   271–276
torsion bar   198
total stress case   75
tourist shuttle. See passenger-vehicle shuttles

tourtia. See glauconitic marl
toxic gases   145
track circuits   151
track systems   162–163
track-to-train radio (TTR)   129–132
tracklaying   170–171
trackwork   16, 157–174
traction motors   177, 179, 184, 186
traffic control and supervision computer (TCSC)   276
traffic levels   157
traffic management. See terminal traffic management (TTM)
trailers   234–235
train captain   175, 179, 190, 194, 200–202, 206, 208, 293
train crew   293
train paths   149, 155, 222, 299
train services   295–296
train servicing   161
train speed   217, 222
Trans-Manche Super-Train Group   176
transition lighting   242
Translink Joint Venture   3
Transmanche Construction   3
Transmanche-Link. See TML
Transport and Road Research Laboratory   91
Travaux du Sud Ouest   164
traverse stations   55
Treaty of Canterbury   6, 301
trunking switch controllers (TSCs)   131
tube à manchette method   36, 111
tunnel boring machines (TBMs)   14–15, 24, 29, 38–40, 48, 57–60, 63, 66–67, 69–73, 75, 80–81, 88–115, 121, 123
turning loops   14
turnouts   159–161, 165–166, 168
TVM 430 signalling system   150–153, 186, 189, 299, 305
twist rails   166

UHF band   132
UIC60 rails   164

UK Fire Research Station reports 232, 234
UK Fire Statistics 231–232, 235
ultimate capacity 292, 304
ultimate-state loading conditions 76
ultraviolet flame detectors 237
UMIST 164
unbalance 135, 137–139
underframe 197, 208
Underground Operations 123–124
Unimog tractor 171
Unitraffic consortium 271
Universal Transverse Mercator projection (UTM) 51
Upper Shakespeare site 86

vans 234–235, 295
vehicle allocation system 258, 276, 278
vehicle management 271, 276–283
ventilation 16, 67–68, 100, 118, 140, 146, 154, 211–216, 238, 242, 247, 249, 291, 294, 303
ventilation shafts 65, 214, 219, 227
versine 173
Verstegan 21

VHF band 132
vibratory sensors 160
viscoelastic relationships 75
visitor access 257
voltage harmonisation 244
von Karman 244
VSB-Stedef 162

wagons 16
walkways 164, 167, 169, 172–174, 189, 194, 241–242, 245, 304
water main 306
Waterloo International 295
Weald and Downland Museum 269
Wealden–Boulonnais dom 26
Westminster Dredging 265
wheelchairs 196
wheelsets 178, 197, 207
wheeltread profile 178
Wimpey 3
Wolff Olins 262
'wriggle' surveys 60

ZED guidance system 57, 98–99
zigzag traverses 56–57